Learning Materials in Biosciences

Learning Materials in Biosciences textbooks compactly and concisely discuss a specific biological, biomedical, biochemical, bioengineering or cell biologic topic. The textbooks in this series are based on lectures for upper-level undergraduates, master's and graduate students, presented and written by authoritative figures in the field at leading universities around the globe.

The titles are organized to guide the reader to a deeper understanding of the concepts covered.

Each textbook provides readers with fundamental insights into the subject and prepares them to independently pursue further thinking and research on the topic. Colored figures, step-by-step protocols and take-home messages offer an accessible approach to learning and understanding.

In addition to being designed to benefit students, Learning Materials textbooks represent a valuable tool for lecturers and teachers, helping them to prepare their own respective coursework.

More information about this series at https://link.springer.com/bookseries/15430

Akatsuki Kimura

Quantitative Biology

A Practical Introduction

 Springer

Akatsuki Kimura
National Institute of Genetics,
and The Graduate University for Advanced Studies (SOKENDAI)
Mishima, Japan

ISSN 2509-6125   ISSN 2509-6133   (electronic)
Learning Materials in Biosciences
ISBN 978-981-16-5017-8   ISBN 978-981-16-5018-5   (eBook)
https://doi.org/10.1007/978-981-16-5018-5

This Springer imprint is published by the registered company Springer Nature Singapore Pte Ltd.
The registered company address is: 152 Beach Road, #21-01/04 Gateway East, Singapore 189721, Singapore

Preface

I wrote this textbook for my younger self, who was then in graduate school. Twenty years ago, I was a student majoring in molecular biology and working at the bench all day. Gradually, I became interested in what is nowadays called quantitative biology. I wanted to calculate the consequences of the models I imagined happening inside the cell. Unfortunately, I did not have the necessary skills in computer programming or the requisite knowledge in physics and mathematics. Fortunately, my supervisor encouraged me to self-study subjects unrelated to my graduate research. As a postdoc, I also had the chance to work in a quantitative biology lab, where I learned from the lab members and continued this self-education. I am continuing to learn, and due to the lack of systematic training, I do not think I am an expert in quantitative biology. However, I think I am an appropriate person for providing beginners with a supportive push toward studying this subject, because I was one of those beginners. I hope that this textbook will be an interesting foundation for a complete beginner to start their journey in the exciting field of quantitative biology.

Another reason for writing this textbook is to discuss cell architectonics. The cell exhibits incredible architecture shaped by nature. Tremendous effort has been made by molecular and cell biologists to understand the structures and functions of cells. We now know that many macromolecules, together with small molecules, make up the cell and regulate its function. However, it is still not fully understood how a large number of tiny (nm scale) molecules assemble and organize themselves to construct a cell on the μm scale that can carry out harmonized functions. This is like asking how a building ~100 m in size (e.g., the Sagrada Familia) can be constructed from building blocks ~10 cm in size, without an architect. This question underlies the general mystery of self-organization found in different fields in general, such as biology, sociology, economics, and medicine. With this textbook, I would like to convince you that quantitative biology methods are necessary for understanding cell architectonics or other self-organization phenomena.

This textbook consists of 11 chapters. Chapter 1 provides an overview of the basic concepts and historical background of quantitative biology. Chapters 2, 3, 7, 9, and 11 present an overview (Chap. 2) of the four major topics (Chaps. 3, 7, 9, and 11) of cell architectonics. These chapters provide a background and purpose for the following chapters on computational methods. Chapters 4, 5, 6, 8, and 10 describe the step-by-step computational methods for beginners. With this combination of chapters describing the biological background and computational methods, I hope that readers will take their first steps toward the rich and deep world of quantitative biology.

Mishima, Japan   Akatsuki Kimura

Acknowledgments

I would like to thank the successive editors of this book, in particular Dr. Misato Kochi, for giving me the opportunity to write. This book is based on what I learned throughout my scientific career, and I would like to thank my mentors, colleagues, students, and collaborators. In particular, Dr. Shuichi Onami, who was my postdoc mentor and gave me a chance to get started in the field, and Dr. Masami Horikoshi, who was my PhD mentor and encouraged me to get into the field. Much of the knowledge contained in this book was provided by my colleagues in QBIO-JP, a Japanese society for quantitative biology. The past and present members of my lab as well as my collaborators played indispensable roles during my pursuit of cell architectonic research. The book is also based on my lectures at the Graduate University for Advanced Studies (SOKENDAI), as well as visiting lectures at Kumamoto University, Osaka University, Keio University, and the University of Tokyo. I thank my colleagues there for giving me the opportunity to lecture and the students for their feedback.

Contents

Introduction to Quantitative Biology

1

Contents

What You Will Learn in This Chapter

In this introductory chapter, we begin by discussing what quantitative biology is. Although the term is used quite often these days, it is difficult to define. In fact, there seems to be no clear-cut and universal definition. We will review historical background leading to interest in the field of quantitative biology. We next discuss why a student or scientist new to the field may want to study quantitative biology using this book. Intended readers of the book are biology students and scientists who are not familiar with quantitative approaches, such as statistical analyses and computer modeling. We hope to convince readers that quantitative ways of thinking and skills are inevitable in modern biology. We will focus on the construction of quantitative models and analyses using computational modeling as a basic skill in quantitative biology. At the end of the chapter, we will discuss the importance of quantitative modeling in quantitative biology and the factors that contribute to a good model.

© Springer Nature Singapore Pte Ltd. 2022   1
A. Kimura, *Quantitative Biology*, Learning Materials in Biosciences,
https://doi.org/10.1007/978-981-16-5018-5_1

Learning Objectives
After completing this chapter, readers should be able to

1. Discuss what quantitative biology is and why it is attracting recent attention.
2. Discuss the role of model construction in biology and why a quantitative approach is necessary.
3. Discuss the properties of a good model.

Important Concepts Discussed in This Chapter

- *Quantitative Biology*; a field of biology emphasizing quantitative approaches, such as data quantification, calculation, or quantitative modeling. Owing to the development of various technologies in biology (e.g., high-throughput genomics or high-resolution imaging), modern biological research has generated extensive quantitative data, and the need for quantitative approaches for data analysis and model construction is increasing.
- *Quantitative Modeling*; the construction of a model of a phenomenon of interest is a major goal in the natural sciences. Qualitative models have been popular in the past; however, as data in the field are becoming more quantitative in nature, quantitative modeling is becoming increasingly important.

1.1   What Is (Modern) Quantitative Biology?

Quantitative biology is a trending area in modern biology. It emphasizes quantitative approaches, such as data quantification, calculation, or quantitative modeling. Because biology, of course, is a branch of the natural sciences, it is hard to imagine a non-quantitative form of biology. There seems no need to mention the famous quote by Galileo Galilei that "*the book of nature is written in the language of mathematics*" (Jesseph 2016); any natural science should be quantitative to some extent.

Then, why do we need the term quantitative biology? A major reason is that there has long been an area of biology that relies mostly on qualitative approaches, which I call here "qualitative biology" for simplicity. Qualitative biology includes classical taxonomy based on qualitative features and gene hunting studies based on qualitative phenotypes. These studies do not require intensive quantification. Qualitative approaches have been important in mainstream biology in the past; accordingly, there was a tendency for students and scientists who are not good at mathematics to study biology. Roughly speaking, mainstream biology has long been a qualitative field.

The situation has been changing, particularly since 2000, when the draft of the human genome was published. For researchers in biology, this meant the end of the era of gene hunting, a typical example of qualitative biology. Before genome projects, the discovery of the nucleic acid sequences of genes whose loss-of-function or ectopic expression led to interesting phenotypes was considered a major contribution. Genome projects have eliminated the need to "discover" gene sequences, which could be obtained within the genome sequence. Uncovering the genotype–phenotype relationship is still an important goal, but efforts to characterize these relationships have decreased dramatically. Biologists, particularly those who are young, have had to think about the new direction in the field, and this has led to the increased interest in quantitative biology.

Systems biology is a term that became popular after 2000. Systems biology aims to obtain a system-level understanding of biological phenomena or processes (Kitano 2000). A system in this case refers to a cell or organism, which can be considered systems of macromolecules (e.g., protein and DNA). To achieve a system-level understanding, classical molecular biology approaches focusing on individual genes are not sufficient. The leaders of systems biology believed that biologists should apply concepts in other fields, such as systems engineering or dynamical systems theory. In addition, genome projects promoted technological advancements, enabling high-throughput (systematic) analyses. Instead of examining the expression of particular genes one-by-one, for example, it has become possible to examine the expression of all genes at once. High-throughput analyses, which are often called "omics" approaches (e.g., genomics, transcriptomics, and proteomics) have provided a massive quantity of data, thereby enabling systems biology approaches. More importantly, this trend in biology has brought in many scientists from outside fields, such as computer sciences and mathematics. In my opinion, systems biology, or post-genome biology, has fostered an environment (i.e., interdisciplinary research) for the emergence of the modern quantitative biology.

One of the earliest and clearest signs of the emergence of modern quantitative biology is the launch of the q-bio (which means quantitative biology) conferences (q-bio.org) in 2007. As indicated on the conference website, "*q-bio... aims at advancing predictive modeling of cellular regulation... The emphasis is on deep theoretical understanding, detailed modeling, and quantitative experimentation directed at understanding the behavior of particular regulatory systems and/or elucidating general principles of cellular information processing.*" The difference between quantitative biology and systems biology might not be clear. Both focus on systems consisting of multiple components (e.g., macromolecules) and are inspired by concepts in systems engineering or dynamical systems theory. Systems biology has a greater focus on big systems (comprehensive, -omics datasets) and the underlying networks, whereas quantitative biology focuses on the dynamics of systems (e.g., time-series analyses) and the scale of analyses may allow the construction of a model to describe the behavior. Quantitative biology also tends to focus on live cell imaging data. According to the website for the Japanese Society for Quantitative Biology (Q-Bio.JP) launched in 2008, "*While most of modern biology was focused on the properties of individual molecules, a future goal will be to understand their dynamics... Now to approach the principles that underlie their dynamical behaviors, the Physical and Chemical Sciences may provide a useful precedent.*" A quantitative approach in biology is not new, nor is the term quantitative biology, as one of the most well-established meetings in biology is named the Cold Spring Harbor Symposia on Quantitative Biology (1933–);

however, modern quantitative biology is *"reaching a new level of maturity as technological advances allow biological systems to be probed and monitored quantitatively with unprecedented control, scope, and resolution"* (Hlavacek et al. 2008).

Questions
1. Provide your own definition of quantitative biology.

1.2   Why Study Quantitative Biology?

Quantitative biology deals with numbers and involves the quantification of results (e.g., digital processing of microscope images and statistical analyses) or the construction of quantitative models (e.g., computer simulations). These processes require skills in mathematics or computer sciences, which are often not taught in ordinary biology classes. Biology research is thought to be conducted by researchers who are not good at math. Do we (biologists) need to teach ourselves the skills for quantitative analyses? Maybe we can collaborate with or leave quantitative analyses to those who excel at math. These are reasonable approaches to save time. However, I strongly recommend that you (biologists) step forward to learn and use those skills in your research, as I did. I was trained as a molecular biologist and ultimately obtained a PhD. At that time, I was not familiar with quantitative skills, as is common in biology. As a postdoctoral researcher, I joined a Systems Biology lab and started to learn how to conduct quantitative research. My mentor and members of the lab taught me many things, but I never received formal training in mathematics or physics. I wish I had. As I cannot go back in time, I decided to continue educating myself, which has been rewarding.

The need for quantitative skills is increasing day by day. It is difficult to publish a paper without statistical or image processing skills. Many biology journals have strict standards regarding the use of statistics. Of course, this is not new, but the demand for rigorous statistical analyses is increasing. For example, Nature journals ask authors to attach a checklist declaring that the appropriate statistical analyses were used. For microscopy images, it is becoming difficult to convince readers of a particular conclusion based on a single representative image. It is often necessary to extract quantitative features from multiple images and demonstrate statistical significance, even for microscope data.

Furthermore, quantification is fundamental to the natural sciences, which is a largely hypothesis-driven field. Richard P. Feynman describes this process concisely in his book (Feynman 2007) as follows: *"In general we look for a new law by the following process. First, we guess it. Then we compute the consequences of the guess to see what would be implied if this law that we guessed is right. Then we compare the result of the computation to nature, with experiment, or experience, compare it directly with observation, to see if it works. If it disagrees with experiment, it is wrong. In that simple statement is the key to science.*" Quantification contributes to each process in hypothesis-driven research. To objectively compute the prediction that follow from a hypothesis, numerical modeling is key. To characterize the behavior of the nature, quantification, such as image processing, is important. To compare observations and evaluate hypotheses, statistical comparisons are needed. Some (or perhaps most) biologists believe that solid results are qualitatively obvious, and results that need to be quantified or tested by statistical methods are trivial. I do not agree with this perspective, as quality research is not limited to areas governed by simple logic. Nature is complicated, as we all know, and laws of nature are often hidden under the complexity. To discover such hidden laws, careful quantitative analyses are necessary. Quantification is also important for another emerging approach, i.e., data-driven science. Data-driven approaches are aimed at extracting knowledge or insights directly from data, without prior hypotheses (Hey 2009). For this approach, computational and statistical skills are clearly necessary.

Biology is becoming an interdisciplinary field, and collaborations among researchers with various backgrounds can benefit from numbers and math, a universal language. For collaboration, it is first necessary to convince others that your biological question is important. To proceed with a collaboration, effective communication is necessary. In addition to the importance of communication for pursuing a specific project, learning concepts in other fields might improve your research. Your important biological questions might be addressed by applying knowledge in other research fields. For example, polymer physics is helpful for understanding chromosome conformation (Mirny 2011) and collective behavior is useful for understanding cytoplasmic streaming (Kimura et al. 2017).

Finally, skills related to quantification are used in fields other than biology, including baseball analytics (Lewis 2003) and finance (Weatherall 2013).

Questions
2. Discuss why quantitative approaches are important for specific fields of biology that you are interested in.
3. Discuss why quantitative approaches were not important in some areas of biology in the past.

1.3   The Aim and Target of This Book

The aim of this book is to enable biologists who are not familiar with computational programming to conduct quantitative analyses and modeling of biological processes. Despite the increasing requirements for quantitative approaches in molecular and cellular biology, the necessary skills have not been integrated into the curriculum, and these approaches are particularly difficult for those who do not have strong backgrounds in math, physics, or computer programming. Many good textbooks written by experts are not aimed at beginners. I was a complete beginner when I started to learn about quantitative approaches in biology and had to teach myself. I wrote this book to help readers who are similar to myself when I started to study quantitative biology. I expect this book to be a comprehensive overview or resource for beginners and encourage readers to continue to learn using advanced textbooks and to incorporate quantitative approaches into their own research.

1.4   Construction of Quantitative Models as a Goal of Quantitative Biology

Quantitative biology covers a wide range of fields within biology and involves diverse approaches. In this textbook, I will focus on the construction of quantitative models and explain elementary strategies and methods related to this goal. The reasons for the focus on the construction of quantitative models are as follows. First, this was my goal when I entered the field of quantitative biology. Accordingly, I believe this might be of interest for many beginners and thus will be a good starting point. Second, as indicated on the q-bio website (*"q-bio . . . aims at advancing predictive modeling of cellular regulation..."*), the construction of quantitative models is a major goal of research in quantitative biology.

1.4.1   What Kind of Model Is a Good Model?

Let me first explain what kind of (quantitative) models we want to construct. "Physical Biology of the Cell" (Phillips et al. 2013), a major quantitative biology textbook, begins with the following sentence (in the preface of the 1st edition). *"With the ever-accelerating pace of scientific knowledge acquisition, particularly in the biological sciences, there is some danger that our scientific understanding of life is becoming a uselessly detailed Borgesian map."* A Borgesian map is a map that appeared in a short story by Jorge Luis Borges and Adolfo Bioy Caseres ("Del Rigor en la Ciencia/On Exactitude in Science," 1946). The Borgesian map is so perfect that its size is as big as the kingdom itself, which is considered useless as a map. Maps are useful when they contain abundant information, but, at the same time, are simple enough to grasp an overview of the land or to determine the appropriate direction. Similarly, scientific knowledge should not only be detailed, but should also be easily understood. Nowadays, scientists produce extensive data, but compared to the production of data, there is relatively little focus on summarizing or extracting essential concepts from data or on effective data presentation. A model in science is a representation of our understanding or hypotheses, just like a map is a representation of information related to land features. The representation can be graphical (as a map), physical, or mathematical. Model construction is a goal in scientific research in which scientists assemble data and test a hypothesis to explain the subject of interest.

What kind of model is a good model? According to the textbook Physical Biology of the Cell, a good model is *"simple enough to be easily grasped by the human mind,"* and at the same time, must *"include at least some of the realistic details...to make reasonably accurate predictions"* (Phillips et al. 2013). This concept is similar to a quote by Albert Einstein, *"Make things as simple as possible, but not simpler"* (Einstein 2010). A model is too simple when it contains limited components or conditions such that the model accounts for limited behaviors or cannot predict behaviors. A model is too detailed when it contains components/conditions that are unnecessary to account for or predict the behavior, or when it is too complicated such that the model becomes a black box, unable to logically connect the causes and consequences of model behavior.

1.4.2 The Need for Quantitative Models

Scientific models, by definition, do not have to be quantitative. A model of DNA duplication via the complementary relationship between adenine-thymidine (A-T) and cytidine-guanosine (G-C) pairs is essentially a qualitative model. When we take into account the binding energy between A-T and C-G pairs, compare the energy between incorrect pairs (e.g., A-C or T-G), and calculate the probability of a mismatch, it becomes a quantitative model. As is evident from this example, there are a number of useful qualitative models, and thus you may question the need for quantitative models. However, qualitative models without quantitative support may be misleading.

In Japan, a popular joke postulates that "if the wind blows, the bucket makers prosper." The rationale ("model") for this statement is as follows. If the wind blows, dust will rise. If dust rises, it will hurt our eyes and the number of optically challenged people will increase. Because a shamisen (three-stringed Japanese lute) player was a typical profession for optically challenged people in the past, demand for shamisen production will rise, and accordingly the demand for cat skin, of which shamisen is made of, will rise. This means that the number of cats will decrease, and the number of rats will increase. Because rats bite buckets (made of wood), the bucket makers will prosper. Each step of the logic may be true; accordingly, the overall statement that "if the wind blows, the bucket makers prosper" might sound true. However, many of us intuitively think that this is not a strong statement. If we quantitatively evaluate each part of the process, we will obtain a probability for the scenario where the bucket makers prosper when the wind blows. A similar story related to the limitation of qualitative logic can be found in the famous Zeno's paradox of the tortoise and Achilles. If we consider the problem in a quantitative manner, we can reach an appropriate, clear conclusion. Therefore, constructing a quantitative model is important to avoid ambiguity, which is obviously important in science.

1.4.3   How Can We Make a Good Quantitative Model?

I hope you are now convinced that quantitative model construction is important in biological sciences. Then, the next question is how are these models constructed? In particular, if we are not familiar with quantitative analyses, what should we do? According to the "Physical Biology of the Cell" textbook (Phillips et al. 2013), there are three important criteria for a good quantitative model in the biological sciences.

First, "*a person who is familiar with a particular biological system writes down a quantitative model for that system*" (Phillips et al. 2013). This person is obviously a biologist who is studying a biological system. Considering that many biologists are not familiar with quantitative modeling, this might sound strange. Rather, one might think that it is efficient and convenient to ask someone familiar with constructing a quantitative model to make one, despite a lack of knowledge about the biological system. However, remembering an appropriate model should be "*as simple as possible, but not simpler,*" suggesting that one has to judge which components should be added or removed from the model. This highly important task in constructing a good model can only be performed by a person who knows the system well, which should be the biologists themselves.

Second, it is necessary to "*incorporate the correct basic physical models*" (Phillips et al. 2013), as living organisms cannot violate the laws of physics. (It should be noted that there might be laws yet to be discovered.) This is again difficult for biologists who are not trained in physics. It may sound impossible to identify and incorporate physical models appropriate for biological systems. The good news is that the number of fundamental physical models important for constructing quantitative models in biology is fairly limited. In the Physical Biology of the Cell textbook, nine theories/models are introduced as fundamental tools to construct quantitative models in biology (Phillips et al. 2013). In my own experience, a relatively small number of physical theories/models are repeatedly used in a wide range of research in the field of quantitative biology. While we (i.e., biologists) need time to learn physics, relatively little effort is required to satisfy the immediate needs for research. In this book, I will only cover a portion of these important theories/models, but I will try to give a good introduction for further studies.

Third, it is necessary to "*use the correct order-of-magnitude estimates for relevant parameters*" (Phillips et al. 2013). This might be the easiest or most familiar condition for most experimental biologists. A straightforward way to fulfill this condition is to perform experiments and measure parameters. However, in some cases, not all relevant parameters are measurable. For example, it is generally difficult to measure the force acting inside the cell. Accordingly, it is important to use estimates that are of the correct order-of-magnitude. The parameters do not have to be measured directly, and the accuracy of the estimation/measurement is not expected to be high. Inside living organisms, most parameters fluctuate. For example, a particular molecule is not always present at a constant number but fluctuates over time. Therefore, a quantitative model that is sensitive to parameter values (i.e., the consequences change dramatically upon small changes in parameter values), the model is unlikely to be applicable to the inside of organisms. In addition, for a general problem, the order-of-magnitude estimates provide a good estimate for the entire system. Such rough calculation is called a Fermi estimation, named after a famous physicist, Enrico Fermi. A typical example of Fermi estimation is the estimation of the number of piano tuners in Chicago based on only the population of the city (~3 million). You might initially be unable to guess, but a rough estimate can actually be obtained. First, the number of families can be estimated to be about 750,000 by assuming that a family contains about four members, on average. The actual family size can be three or five, but the important point is that the order of magnitude is correct. An average family should contain between one and ten members. Second, you can guess that the number of pianos is about 150,000 by guessing that one out of five families will own a piano. Third, guessing that a piano tuner takes ~2 h to tune a piano, and he/she works 8 h/day, 5 days/week, 50 weeks/year, the piano tuner can tune 1000 pianos per year. Finally, by guessing that a piano is tuned once per year on average, the number of piano tuners is 150 (150,000 pianos divided by 1000). This kind of rough estimation, or back-of-the-envelope calculation, often provides a surprisingly accurate estimation. This is generally true for quantitative models in biology. Often, order-of-magnitude estimates are sufficient to develop appropriate models.

In the following chapters, I will provide a basis for true beginners to construct quantitative models.

Questions
4. Describe the major challenges you face when employing quantitative approaches.
5. Discuss the role of model construction in biology, and why a quantitative approach is essential.
6. Discuss the criteria for a good model.

Answers
1–6. There is no single answer to the questions. Please consider yourself, and discuss with your colleagues if possible.

Take-Home Message
- Quantitative biology is an emerging field in modern biology. It emphasizes quantitative approaches, but a clear-cut or universal definition has not been established.

(continued)

- Quantitative biology is becoming progressively more important; modern biology is full of quantitative data and the need for quantitative approaches to evaluating data is increasing.
- Construction of quantitative models is a major goal of quantitative biology. With such models, we can explain and predict biological phenomena in a quantitative manner.
- Good models should be simple enough to focus on essential components, and complex enough to explain and predict the essential aspects of phenomena.

References

Einstein A. The ultimate quotable Einstein. Princeton University Press; 2010.

Feynman RP. The character of physical law. London: Penguin; 2007.

Hey AJG. The fourth paradigm. Microsoft Research; 2009.

Hlavacek WS, Edwards JS, Jiang Y, Nemenman I, Wall ME, Faeder JR. Editorial: selected papers from the first q-bio conference on cellular information processing. IET Syst Biol. 2008;2:203–5.

Jesseph D. Ratios, quotients, and the language of nature. The language of nature. University of Minnesota Press; 2016. p. 160–77.

Kimura K, Mamane A, Sasaki T, et al. Endoplasmic-reticulum-mediated microtubule alignment governs cytoplasmic streaming. Nat Cell Biol. 2017;19:399–406.

Kitano H. Perspectives on systems biology. New Gen Comput. 2000;18:199–216.

Lewis M. Moneyball: the art of winning an unfair game. W W Norton & Co Inc; 2003.

Mirny LA. The fractal globule as a model of chromatin architecture in the cell. Chromosome Res. 2011;19:37–51.

Phillips R, Kondev J, Theriot J, Orme N. Physical biology of the cell. Garland Pub; 2013.

Weatherall JO. The physics of wall street: a brief history of predicting the unpredictable. Houghton Mifflin Harcourt; 2013.

Further Reading

Milo R, Phillips R. Cell biology by the numbers. Garland Science; 2015.

Cell Architectonics

2

Contents

What You Will Learn in This Chapter

Quantitative biology combines several scientific fields to study biological systems. While I expect that the concepts and methodologies in this textbook will be applicable to many topics, I decided to focus on a particular subject—the spatial organization and dynamics of cells; this is mainly because I am working in this field. In addition, this topic raises various interesting questions common to other fields related to quantitative biology, such as mechanical modeling, diversity, self-organization, and development over time. In this chapter, I will introduce cell architectonics, which I work on, and the four main objectives of the field. You can skip this chapter if you are not interested in this particular subject; however, I believe this chapter can help orient readers for understanding the following chapters.

© Springer Nature Singapore Pte Ltd. 2022

A. Kimura, *Quantitative Biology*, Learning Materials in Biosciences,

https://doi.org/10.1007/978-981-16-5018-5_2

2.1   Why We Deal with the Architectonics of the Cell (In This Book)?

Quantitative biology is a methodology-based research field. Roughly speaking, any biological research involving quantitative approaches can be categorized as quantitative biology. Therefore, as the possible subjects of quantitative biology are unlimited, I hope that this textbook will be beneficial for learners interested in various biological topics.

Meanwhile, as a researcher, I am interested in particular research subjects, and thus my viewpoint is biased. Examples and exercises in this book will also be biased. Therefore, I think it is necessary to clarify my research interests for the readers, and how they are related to quantitative biology.

My research interests include understanding how the cell is constructed from macromolecules (i.e., gene products). I call this field of research the architectonics of the cell or cell architectonics. Because my research subject is the cell, cell architectonics falls within the field of cell biology. There are several features that I think are important for understanding cell architectonics, which I will briefly explain in this chapter and in more detail in other chapters in this book.

Quantitative biology approaches are necessary for the field of cell architectonics. This is because a major purpose of cell architectonics is understanding cellular mechanics. For this purpose, structural calculations of the cell are required, similar to what architects must do when designing buildings. Developing quantitative models of the cell is warranted for performing cell structural calculations. Therefore, quantitative biology has a high potential as a research field for elucidating the architecture of the cell.

2.2   What Is Cell Architectonics?

I classify "cell architectonics" as a subfield of cellular biology that aims to understand how spatial organization of the cell is achieved. However, this term is not popular, and I am almost alone in using it. As the questions cell architectonics aims to answer are popular, other researchers think these questions are typical in cell biology and that an additional term is not needed. I personally think it is still meaningful to use cell architectonics because it emphasizes an important frontier in cell biology that requires interdisciplinary efforts between biology (genetics and molecular biology) and physics (mechanics and self-organization).

Within cell architectonics, I consider there to be four main objectives: understanding the (1) mechanics, (2) diversity, (3) self-organization, and (4) time-dependent development of the cell. These concepts are briefly explained in the following sections, and are described in more detail in other chapters focusing on each objective.

2.3   Objective #1: Mechanics of the Cell (Chap. 3)

The mechanical understanding of cells is a fundamental objective of cell architectonics. As are structural calculations of buildings inevitable in architectonics, structural calculations of the cell are also important for understanding the architecture of the cell.

In the structural calculations of buildings, the major players are the physical forces and mechanical properties of the materials. Similarly, in cell architectonics, the physical forces and mechanical properties are critical. The cytoskeleton and motor proteins are well-known sources of physical forces inside the cell. The cytoskeleton also plays a key role in determining the mechanical properties of cellular components. Thanks to the efforts in the field of biophysics, there is increasing knowledge on the force production and mechanical properties of cellular materials when they are isolated from cells (in vitro). However, determining the force production and mechanical properties in vivo is still challenging. Therefore, structural calculation (or quantitative modeling) of the cell requires compromises between accuracy and simplicity, as discussed in Chap. 1.

2.4   Objective #2: Diversity of the Cell (Chap. 7)

Cellular appearance differs from one cell to the next. This diversity is why an analogy between cells and buildings/cities, rather than machines, is appropriate. Buildings and cities also differ from one another, although their materials and functions are similar. These differences result from the surrounding environment and other factors, such as culture. In contrast, machines (industrial products) that perform the same function must be identical. Within this context, the cell is not a machine, but a building/city.

Some aspects of cellular diversity can be explained by the same model with different parameters. Therefore, my expectation is that by constructing an appropriate model for the cell, we can explain and understand cellular diversity only by changing the values of the parameters in the model. We have applied this approach to understand cellular diversity, as explained in Chap. 7.

2.5   Objective #3: Self-Organization of the Cell (Chap. 9)

When I was a freshman at university, I was interested in architectonics, more specifically, in city planning. At that time, I did not imagine that I would become a cell biologist. One important trigger that led me to cell biology was a book written by an architect, Yoshinobu Ashihara. In the book *The Hidden Order* (Ashihara 1989), Ashihara discusses how cities in Japan, such as Tokyo, emerged spontaneously. The cities have an astonishing degree of spatial order and harmony, even though the cities lack a rigid and global design plan and the citizens act selfishly (Fig. 2.1). Such organization is not designed beforehand or intentionally, and therefore Ashihara described this phenomenon as a "hidden order." I

Paris   Tokyo

Fig. 2.1 Cities with (Paris) and without (Tokyo) a clear top-down design. These photos of Paris and Tokyo were obtained from Google Earth at a similar scale. This image is reprinted with a modification and permission from a web article on cell architectonics (https://gendai.ismedia.jp/articles/-/55032? page=2). Map data: Google (Paris) and ZENRIN (Tokyo)

thought that this hidden order, or so-called self-organization, is everywhere around us and worth studying.

Indeed, the cell also appears to have a hidden order. The cell is made up of various macromolecules, but none of these molecules is the leader, designer, or architect that is aware of the overall order and can tell the other molecules what to do. Instead, each molecule performs its own job. The genome is often described as the blueprint of a cell or organism. However, the information encoded in the genome is how to make RNA, without direct information about the order or organization. Despite the absence of a leader, the molecules organize themselves to construct an ordered cell. My younger self thought that the cell was the best subject to study the mystery of self-organization, and thus I decided to become a cell biologist.

Understanding self-organization is thus a major objective of my version of cell architectonics. I hope to understand how a population of molecules assembles into a cell that achieves unified functions. For this purpose, numerical modeling is a powerful tool. For example, agent-based modeling (ABM) can model the behavior of a system (e.g., a cell) consisting of many active components (e.g., molecules) that act according to their own rules. Using ABM, we can ask what kind of rules for each molecule reproduces a certain cellular behavior. Thus, quantitative modeling is an effective approach for this purpose.

2.6   Objective #4: Development of the Cell over Time (Chap. 11)

Self-organization during cellular construction involves many unanswered questions that should be addressed in future research. At the same time, self-organization is a well-accepted concept, with some elucidated mechanisms. I would like to state that the true frontier in cell architectonics, or biology in general, is understanding the development/ transition of self-organized architecture over time. While the status of the cell at each time point is accomplished in a self-organized manner, the cells or organisms often alter their status. These changes are sometimes very drastic, such that the order created previously is almost completely destroyed, followed by the creation of a different order. Such a transition is often explained by temporal changes in gene expression, as a genetic program encoded in the genome. However, to what extent does the genetic program alone explain these robust and dynamic transitions? Therefore, we need to uncover how the transition between self-organized states is organized. Even though this is an uncultivated research area in biology, I believe these knowledge gaps cannot be filled without the help of quantitative modeling approaches.

Take-Home Message
- "Cell architectonics" is a subfield of cell biology that aims to understand how cellular spatial organization is achieved. However, this is not an established field.
- The author of this book believes that it is meaningful to distinguish cell architectonics from conventional cell biology to focus on the mechanics, diversity, self-organization, and development over time of the spatial organization of the cell.
- Cell architectonics shares concepts and methodologies with quantitative biology.

Reference

Ashihara Y. The hidden order: Tokyo through the twentieth century. New York: Kodansha USA Inc; 1989.

Further Reading

Kimura A. Introduction to cell architectonics (Japanese). Kogakusha; 2019.

Mechanics of the Cell

3

Contents

> **What You Will Learn in This Chapter**
>
> The mechanics of the cell is a typical subject in quantitative biology. It is also a critical topic in cell architectonics. The movement of substances within the cell and deformation of the cell are caused by mechanical forces. However, it is difficult to directly measure the force generated inside a cell. Therefore, understanding cellular mechanics is challenging. Currently, we can only gather indirect pieces of evidence and combine them to understand cell mechanics. This often requires multidisciplinary approaches. The central methods, such as measuring forces generated by molecules or measuring the physical properties of intracellular structures, rely on biophysics. Cell biology approaches are also necessary, for example, to describe

(continued)

© Springer Nature Singapore Pte Ltd. 2022   17
A. Kimura, *Quantitative Biology*, Learning Materials in Biosciences,
https://doi.org/10.1007/978-981-16-5018-5_3

intracellular phenomena based on live cell imaging. In this chapter, I will introduce the properties of the cell's mechanical elements, which include the cytoskeleton, cell membrane, and cytoplasm.

Learning Objectives
After completing this chapter, readers should be able to

1. Understand what Reynolds number is, and that viscosity dominates over inertia inside the cell.
2. Understand that most cellular materials contain both elastic and viscous properties, depending on the time and length scale of the deformation. This property is called viscoelasticity.

Important Concepts Discussed in This Chapter

* *Reynolds number*; quantity describing the ratio between inertia and viscosity of a system. When this number is low, viscosity dominates over inertia, as seen in most cellular phenomena.
* *Viscoelasticity*; most cellular materials contain both elastic and viscous properties and thus have viscoelastic properties.

3.1   Mechanical Forces and Cellular Dynamics

The cell's architecture is not static but is very dynamic in that the shape of the cell or organelle and the positions of the inside components constantly change (Fig. 3.1). As mechanical forces are required to change the shape or move objects around us, mechanical forces are also required to change the shape and positions inside the cell. However, it is difficult to accurately measure the force generated by a cell. Cells are surrounded with layers of plasma membrane and cell walls, and thus the measurement apparatuses cannot reach inside the cells in most cases. There are effective methods for manipulating force probes, such as optical or magnetic tweezers, from remote sites; however, the probes are still required to get inside the cell, which is invasive to some extent.

3.2   Methods for Applying Force to Cellular Materials

Atomic force microscopy (AFM) is a technique for measuring shapes by scanning the surface of an object with a cantilever probe. AFM can also measure the repulsive force by pushing the tip of the probe into the surface with an arbitrary force.

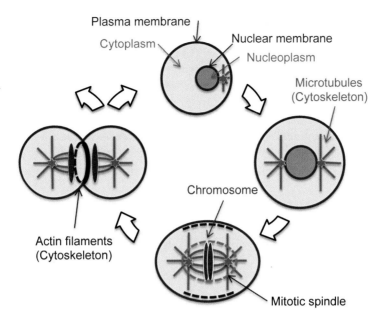

Fig. 3.1 Schematic diagram of mechanical elements and the changes they undergo during the cell cycle. Immediately after division, the cells and nuclei are small, and the nuclei are often not located in the center (top). As the cell cycle progresses, the nucleus size and cell volume increase, and the nucleus moves to the center of the cell (right). During mitosis, the nuclear envelope collapses, forming a spindle consisting of condensed chromosomes and microtubules in the center of the cell (bottom). In addition, actin fibers begin to accumulate in the division plane. Cytokinesis occurs when the ring structure of actin fibers accumulates on the division plane and constricts the cell. Chromosome segregation occurs by moving sister chromatids to the poles of the mitotic spindle (left). This produces new daughter cells consisting of chromosomes and cytoplasm (top). The figure is reprinted after modification from Kimura (2016)

Laser irradiation is used to generate localized heat at a focused position inside a cell where the cell structures are destroyed only at that location. By measuring the movements of objects after laser irradiation, we can estimate the tension before irradiation (Grill et al. 2001; Mayer et al. 2010; Ishihara and Sugimura 2012).

It is also possible to apply forces to cells by confining them to micro-patterns. In the case of cells that do not adhere to a substrate, such as embryos, they can be transformed into any shape by fitting the cells into a mold made of polydimethylsiloxane (PDMS) resin (Minc et al. 2011; Chang et al. 2014). For cells that adhere to a substrate, such as HeLa cells, their shape can be controlled by printing the molecule that serves as the substrate in an arbitrary shape (Théry et al. 2007).

3.3   Mechanical Properties of Structures Inside the Cell

Even if we succeed in quantifying the forces generated inside the cell, the information obtained is not sufficient. The changes caused by these forces depend on the mechanical properties of the objects which the forces act upon. If the object is soft, the forces may change its shape but not move it. In contrast, if the object is rigid, the object may move without any deformation. Therefore, to understand the mechanics of the cell, we need to know both the forces generated inside the cell and the mechanical properties of the objects being acted upon. It is also difficult to directly determine the mechanical properties of cellular structures inside the cell (in vivo); thus, we can characterize them by collecting and combining indirect information. Theories of physics are important for combining multiple measurements to obtain a unified view of the mechanical properties of cellular structures.

To simplify the complicated environment inside the cell to obtain a rough model connecting cell biology and physics, I propose classifying the mechanical elements in the cell according to the characteristics of their shape (Fig. 3.2). One-dimensional structures include the cytoskeleton (e.g., microtubules and actin filaments) and chromosomes. Two-dimensional structures include the plasma membrane and nuclear envelope, while three-dimensional structures include the cytoplasm and nucleoplasm. We can then utilize this simplification for applying theories. This includes theories of elastic rods or polymers for one-dimensional structures, surface theory for two-dimensional structures, and fluid dynamics for three-dimensional structures. Because the actual structures of these components are complicated, it is up to the researchers to decide which theory/model is suitable for each structure.

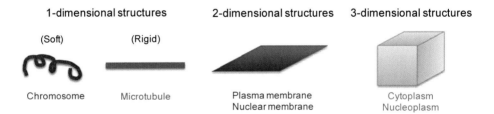

Fig. 3.2 Classification of cellular components based on their shape. An example of applying theories of physics to the simplified classification of cellular components. Here, mechanical elements are roughly classified as one-dimensional, two-dimensional, or three-dimensional structures. The figure is reprinted after modification from Kimura (2016)

3.4   Relationship Between Intracellular Deformation and Force: Elasticity, Viscosity, and Viscoelasticity

When a force is applied to an object, the object to some extent resists moving or deforming. For example, if a force is applied to a straight cytoskeleton causing it to bend, the cytoskeleton attempts to return to its straight state. To bend further, more force must be applied. When strain and stress are in proportion, this object is said to have elasticity, which is generally a property of solid objects. In another case, the object may not return to its original shape, even if the force is released. This is seen for liquids, as they do not return to their original position after being moved by a force; this property is called viscosity. A solid material can also exhibit viscosity if its original shape is not restored after releasing a force, which was continuously applied for a sufficiently long time to cause deformation as well as rearrangement of the inside components.

Objects inside a cell can also be viscoelastic, i.e., bodies with both elastic and viscous properties. For example, the mitotic spindle is a piece of intracellular machinery that segregates sister chromosomes into daughter cells upon mitosis. When it is deformed by applying an external force, mitotic spindles behave like an elastic or viscous body according to the speed and duration of the force (Itabashi et al. 2009; Shimamoto et al. 2011). The properties of elasticity or viscosity quantitatively depend on the time scale and the magnitude of deformation.

3.5   Stress–Strain Relationship of Elastic Materials

Let us consider the relationship between the force (stress) and deformation (strain) when elasticity is the main property. Deformation (strain) includes tension, compression, shear, torsion, and bending. Substances generally maintain their original shape and generate a resistance force (Fig. 3.3). Because it is necessary to apply a large force for large deformations, the amount of force applied inside the cell can be estimated from the degree of deformation. To estimate the amount of force from the deformation, it is necessary to know the stress–strain relationship, which is the relationship between the amount of force required to deform a substance to a certain degree.

Strain and stress can often be approximated as proportional. To understand this, let us consider a spring, which is a typical elastic material. When a one-dimensional spring is pulled or compressed for elongation or shortening, respectively, the external force required to change the original length is proportional to the degree of deformation. This relationship is similar for two-dimensional film structures and three-dimensional objects. Such a model can explain the cell arrangement pattern of epithelial cells (Farhadifar et al. 2007). Similarly, the bending of one-dimensional microtubules or two-dimensional cell membranes can be modeled based on the notion that the force is proportional to the bending distance (Howard 2001; Deuling and Helfrich 1976). I would like to introduce the buckling phenomenon as well, which is another characteristic of elastic bodies. In the

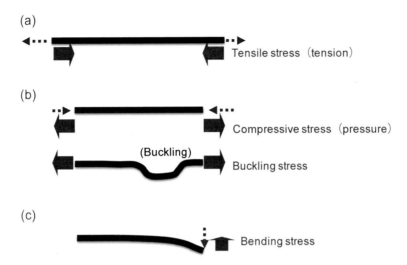

Fig. 3.3 Deformation (strain) and stress of mechanical elements. (**a–c**) Diagram illustrating the stress for three types of strain. (**a**) Expansion. (**b**) Compression. (**c**) Bending. The figure is reprinted after modification from Kimura (2016)

case of a thin rod-shaped or film-like structure, when a force is applied in the direction of compression, the length of the structure is maintained and instead it is deformed while the distance between the end points is shortened. In the case of a rod-shaped structure, the force required for buckling is inversely proportional to the square of the length of the structure (i.e., the longer the structure, the easier it is to buckle) (Howard 2001).

Measuring the relationship between stress and strain in vivo is difficult. A popular strategy is to first characterize the stress-strain relationship outside the cell (i.e., in vitro), and then estimate the force from the deformation observed in vivo. For example, the Xenopus cell-free extract system is a powerful in vitro system that enables force measurements in an environment that is similar to that in vivo.

In addition to deformation by extrinsic forces, biological materials actively deform themselves via polymerization and depolymerization. Such active deformation can produce forces. For instance, the elongation and shortening of microtubules can push or pull components attached to their ends.

The theories of elastic materials can explain the spatial patterning observed in organisms. One example is an elegant study on the retinal pattern formation of the fruit fly, *Drosophila melanogaster* (Hayashi and Carthew 2004). The authors of this study demonstrated that the various patterns observed in wild-type and mutant strains can be explained by the theory of surface mechanics considering the balance of surface tension. Moreover, in our research, we applied the theory of bending elasticity to explain the dynamic changes in cell shape during cytokinesis (Koyama et al. 2012).

3.6   Rheology

Rheology, a discipline that deals with deformation and the flow of matter, can be applied to complex components, such as the cytoplasm, to address its physical properties. To analyze a fluid, passive rheology, which observes the passive movement of a probe floating in the fluid, can be used to infer the properties of the fluid. Passive rheology of the cytoplasm is performed by injecting a foreign substance (e.g., artificial beads) into the cell or tracking its trajectory using the intracellular structure as a probe. On the other hand, the active rheological approach employs probes that are actively moved to record the responses of the surrounding fluid. Probes include polystyrene beads that are injected and moved with optical tweezers (Guo et al. 2014) or magnetic beads that are moved with magnetic tweezers (Garzon-Coral et al. 2016).

3.7   Reynolds Number

In high-school mechanics, you might have learned Newton's equation of motion $\mathbf{F} = m\mathbf{a}$, which states that if force $\mathbf{F}$ is applied to an object with mass m, the object moves with acceleration $\mathbf{a}$. Likewise, you might have learned that the above equation holds true for an ideal case in the absence of drag, whose force is usually proportional to the object velocity. Hence, considering drag, the force balance is given by $\mathbf{F} = m\mathbf{a} + k\mathbf{v}$, where $\mathbf{v}$ is the object velocity and k is the drag coefficient. Consequently, when the drag coefficient k is large, the applied force $\mathbf{F}$ moves the object at an almost constant speed with low acceleration under a regime known as viscous dominant. In contrast, if the drag coefficient k is small, force $\mathbf{F}$ accelerates the object under an inertia-dominant regime.

   The Reynolds number (Re) (Purcell 1977) is a good measure for evaluating whether a regime is viscous or inertia dominant and is defined as:

$$Re = \frac{\rho s l}{\eta} \tag{3.1}$$

where ρ is the density of the fluid surrounding the object, s is the speed of the fluid with respect to the object, l is the characteristic length, and η is the dynamic viscosity of the fluid.

   Let us calculate the Reynolds number for nuclear migration in the *Caenorhabditis elegans* embryo. Density ρ should be that of the cytoplasm, which is almost equal to the density of water, and thus $\rho \approx 10^3$ kg/m^3. Speed s of nuclear migration is approximately 0.1 μm/s, or $s \approx 10^{-7}$ m/s. The characteristic length can be considered equal to the cell length at $l \approx 10^{-5}$ m (i.e., 10 μm). Note that we are only considering the order of magnitude, and thus rough approximations are appropriate. The dynamic viscosity of the cytoplasm is $\eta \approx 1$ Pa s (Daniels et al. 2006), which is approximately 100 times that of water. Using these values, the Reynolds number for nuclear migration in the cell is approximately 10^{-9}. A Reynolds number below 1 indicates a viscous-dominant regime, whereas a Reynolds number above 1 indicates an inertia-dominant regime. Therefore, $Re = 10^{-9}$ indicates that the cell exhibits a viscous-dominant regime, and inertia can be neglected. In fact, inertia can be generally neglected when analyzing movements inside the cell because the speed s and characteristic length l are small. This means that if the velocity of an object can be obtained from microscopic observation, it can be estimated that the object is subjected to a force proportional to the velocity.

> **Questions**
> 1. Calculate the Reynolds number for the action of throwing a ball. Let the density of air be 1.2 kg/m^3, the speed of the ball 50 km/h, the size of the ball 10 cm, and the dynamic viscosity of air 2×10^{-6} Pa s.
> 2. Determine whether the ball movement in question #1 is inertia- or viscosity-dominant.

3.8   Equations for Describing Viscous Fluids

The Navier-Stokes equation is a commonly used equation for describing the behavior of a fluid. The equation itself is complicated, but for intracellular phenomena, incompressible (constant density) and constant viscosity can be assumed. In addition, because the Reynolds number is low, the inertial force can be ignored in most cases. Thus, the equation for a fluid under these conditions can be relatively simplified as follows:

$$-\nabla P(x, y, z) + \mu \nabla^2 \vec{v}(x, y, z) + \vec{f}(x, y, z) = \vec{0} \tag{3.2}$$

$$\nabla \cdot \vec{v}(x, y, z) = 0 \tag{3.3}$$

where ∇ is the operator for finding the gradient, (x, y, z) is the position, P is the pressure, μ is the absolute viscosity, v is the flow velocity, and f is the external force. We have shown that equations representing such simple viscous fluids can describe the cytoplasmic streaming that occurs in *C. elegans* and mouse embryos (Niwayama et al. 2011, 2016).

3.9   Modeling Cell Behaviors Based on Cell Mechanics

Theories describing the mechanics of cellular materials provide basic assumptions for modeling the architecture of a cell. In my own research, we studied topics such as how the nucleus finds the center of a cell (Kimura and Onami 2005, 2007; Tanimoto et al. 2016; Kimura and Kimura 2011), how cells are positioned inside the eggshell of an embryo (Yamamoto and Kimura 2017), and how cytoplasmic streaming is generated (Niwayama et al. 2011; Kimura et al. 2017). In these studies, we applied the physical theory of elasticity, surface tension, and fluid dynamics to connect the function of the molecules to the behaviors of the cell. As a biologist, one might think that studying physics to model biological behavior is a huge task. However, "*a relatively small number of fundamental physical model can serve as the foundation of whole bodies of quantitative biological intuition, broadly useful across a wide range of apparently unrelated biological problems*" (p. viii of Phillips et al. 2013). In the next chapter, we will create a simple but biologically relevant model to experience model construction procedures.

Answers

1. Approximately 7×10^5. Using Eq. (3.1), the Reynolds number is calculated as $Re = \rho s l / \eta$. In the current example, $\rho = 1.2$ kg/m^3, $s = 50$ km/h $= 1.4$ m/s, $l = 10$ cm $= 0.1$ m, and $\eta = 2 \times 10^{-6}$ Pa s.
2. Inertia dominant. If the Reynolds number exceeds 1, the inertia is dominant. This result is consistent with our observation that a ball thrown in the air continues to move after it leaves our hands.

Take-Home Message

- Physical laws in mechanics can be applied for modeling the spatial organization of the cell.
- The appropriate laws of physics depend on the length and time scale of the phenomenon.
- The Reynolds number is an important measure for describing the cellular environment. If the focus is on small movements within viscous fluids, which is the case for most cellular movements, the Reynolds number is small, and consequently inertia can be neglected.
- Viscoelasticity is another important concept for understanding cell mechanics. The same material can exhibit both elasticity and viscosity. The dominant property depends on the length and time scales of the behavior.

References

Chang F, Atilgan E, Burgess D, Minc N. Manipulating cell shape by placing cells into microfabricated chambers. Methods Mol Biol. 2014;1136:281–90.

Daniels BR, Masi BC, Wirtz D. Probing single-cell micromechanics in vivo: the microrheology of *C. elegans* developing embryos. Biophys J. 2006;90:4712–9.

Deuling HJ, Helfrich W. Red blood cell shapes as explained on the basis of curvature elasticity. Biophys J. 1976;16:861–8.

Farhadifar R, Röper J-C, Aigouy B, Eaton S, Jülicher F. The influence of cell mechanics, cell-cell interactions, and proliferation on epithelial packing. Curr Biol. 2007;17:2095–104.

Garzon-Coral C, Fantana HA, Howard J. A force-generating machinery maintains the spindle at the cell center during mitosis. Science. 2016;352:1124–7.

Grill SW, Gönczy P, Stelzer EH, Hyman AA. Polarity controls forces governing asymmetric spindle positioning in the *Caenorhabditis elegans* embryo. Nature. 2001;409:630–3.

Guo M, Ehrlicher AJ, Jensen MH, Renz M, Moore JR, Goldman RD, Lippincott-Schwartz J, Mackintosh FC, Weitz DA. Probing the stochastic, motor-driven properties of the cytoplasm using force spectrum microscopy. Cell. 2014;158:822–32.

Hayashi T, Carthew RW. Surface mechanics mediate pattern formation in the developing retina. Nature. 2004;431:647–52.

Howard J. Mechanics of motor proteins and the cytoskeleton. Sinauer Associates, Inc; 2001.

Ishihara S, Sugimura K. Bayesian inference of force dynamics during morphogenesis. J Theor Biol. 2012;313:201–11.

Itabashi T, Takagi J, Shimamoto Y, Onoe H, Kuwana K, Shimoyama I, Gaetz J, Kapoor TM, Ishiwata S. Probing the mechanical architecture of the vertebrate meiotic spindle. Nat Methods. 2009;6: 167–72.

Kimura A. Quantitative biology of cell division (Japanese). In: Kobayashi T, editor. Quantitative biology, Dojin Bioscience Series. Kyoto: Kagaku-Dojin; 2016. p. 56–69.

Kimura K, Kimura A. Intracellular organelles mediate cytoplasmic pulling force for centrosome centration in the *Caenorhabditis elegans* early embryo. Proc Natl Acad Sci U S A. 2011;108:137– 42.

Kimura A, Onami S. Computer simulations and image processing reveal length-dependent pulling force as the primary mechanism for *C. elegans* male pronuclear migration. Dev Cell. 2005;8:765– 75.

Kimura A, Onami S. Local cortical pulling-force repression switches centrosomal centration and posterior displacement in *C. elegans*. J Cell Biol. 2007;179:1347–54.

Kimura K, Mamane A, Sasaki T, Sato K, Takagi J, Niwayama R, Hufnagel L, Shimamoto Y, Joanny JF, Uchida S, Kimura A. Endoplasmic-reticulum-mediated microtubule alignment governs cyto-plasmic streaming. Nat Cell Biol. 2017;19:399–406.

Koyama H, Umeda T, Nakamura K, Higuchi T, Kimura A. A high-resolution shape fitting and simulation demonstrated equatorial cell surface softening during cytokinesis and its promotive role in cytokinesis. PLoS One. 2012;7:e31607.

Mayer M, Depken M, Bois JS, Jülicher F, Grill SW. Anisotropies in cortical tension reveal the physical basis of polarizing cortical flows. Nature. 2010;467:617–21.

Minc N, Burgess D, Chang F. Influence of cell geometry on division-plane positioning. Cell. 2011;144:414–26.

Niwayama R, Shinohara K, Kimura A. Hydrodynamic property of the cytoplasm is sufficient to mediate cytoplasmic streaming in the *Caenorhabditis elegans* embryo. Proc Natl Acad Sci U S A. 2011;108:11900–5.

Niwayama R, Nagao H, Kitajima TS, Hufnagel L, Shinohara K, Hi-guchi T, Ishikawa T, Kimura A. Bayesian inference of forces causing cytoplasmic streaming in *Caenorhabditis elegans* embryos and mouse oocytes. PLoS One. 2016;11:e0159917–8.

Phillips R, Kondev J, Theriot J, Orme N. Physical biology of the cell. Garland Pub; 2013.

Purcell EM. Life at low Reynolds number. Am J Phys. 1977;45:3–11.

Shimamoto Y, Maeda YT, Ishiwata S, Libchaber AJ, Kapoor TM. Insights into the micromechanical properties of the metaphase spindle. Cell. 2011;145:1062–74.

Tanimoto H, Kimura A, Minc N. Shape–motion relationships of centering microtubule asters. J Cell Biol. 2016;212:777–87.

Théry M, Jiménez-Dalmaroni A, Racine V, Bornens M, Jülicher F. Experimental and theoretical study of mitotic spindle orientation. Nature. 2007;447:493–6.

Yamamoto K, Kimura A. An asymmetric attraction model for the diversity and robustness of cell arrangement in nematodes. Development. 2017;144:4437–49.

Further Reading

Forgacs G, Newman SA. Biological physics of the developing embryo. Cambridge University Press; 2005.

Boal D. Mechanics of the cell. Cambridge University Press; 2002.

Implementing Toy Models in Microsoft Excel

4

Contents

> **What You Will Learn in This Chapter**
>
> This chapter introduces two important concepts to biologists willing to focus on research using quantitative approaches. First, it illustrates how to evolve from hypotheses and descriptive principles to mathematical formulations. Second, it provides a step-by-step introduction to simulate simple models using Microsoft Excel. I will introduce a biological process, namely, the movement of the nucleus within a cell, as the example to develop our model. Specifically, we will construct a simple, one-dimensional model relying on minimal knowledge of cell mechanics. Then, we will implement the model using Microsoft Excel, a spreadsheet software, which should be familiar to most readers. As Microsoft Excel is familiar, I believe that conducting a simulation in this software will provide a straightforward starting

(continued)

© Springer Nature Singapore Pte Ltd. 2022
A. Kimura, *Quantitative Biology*, Learning Materials in Biosciences,
https://doi.org/10.1007/978-981-16-5018-5_4

point. This implementation represents a preparation step for the next chapter, where we will follow the same procedure but using Python, which enables more comprehensive modeling. This chapter aims to provide basic tools for biologists to start exploring simple quantitative models. Furthermore, the chapter intends to demonstrate that simulations and mathematical models are not as complicated as they might seem at firsthand. Therefore, this chapter is expected to lay the foundation and motivation for the deeper exploration of quantitative approaches.

Learning Objectives
After completing this chapter, readers should be able to

1. Express a simple hypothesis as a set of numerical relationships among different components.
2. Implement the numerical relationships in a spreadsheet software, such as Microsoft Excel, and evaluate the hypothesis.

Important Concepts Discussed in This Chapter

- *Stokes' law*; relationship between the force applied to an object and its velocity in a fluid when inertia can be neglected.
- *Buckling force*; force required to buckle an elastic rod.

4.1 Custom Makes All Things Easy

For a beginner, there are many ways to learn quantitative biology. A royal road is to study basic mathematics, physics, or computer programming. This is certainly recommended for young students who have the required energy, time, and patience. As mentioned before, I myself started studying quantitative methods after getting a PhD and enrolled as a post-doc. I was not supposed to focus solely on studying, but also must conduct research as a post-doc. Thus, I started to construct quantitative models and acquired the knowledge and methodologies according to my needs. With this approach, I was able to move my research forward while learning new things. This approach was also effective to keep myself motivated to study, because I am often not patient enough to go through lengthy textbooks.

I will adhere to this approach and start from the model construction, avoiding intricate details, and explain the necessary physics or programming concepts whenever needed. I hope this approach, despite not being a royal road, would be effective for many busy biologists to get started with quantitative approaches.

4.2 The Toy Model: Centration of the Nucleus Inside a Cell

As toy model to introduce quantitative methods, I chose the problem of how the nucleus reaches and stays at the center of the cell. Actually, this is exactly the same model I first constructed when I changed my field of research to quantitative biology. Furthermore, I find this problem to be a very good starting point, as it is simple and allows to grasp different aspects of quantitative biology.

4.2.1 Biological Background

If asked to draw a cell, I believe many of you draw a circle in the center of a larger circle or square to depict the nucleus within a cell. This popular view of the nucleus located at the center of the cell is correct in most cases, only with some exceptions. Then, how does the nucleus identify the center of the cell and position itself there? Biologists got interested in this problem soon after the invention of the microscope, when they saw the dynamic movement of the sperm-derived nucleus moving toward the center of the egg after fertilization (Wilson 1925). Interestingly, this movement was almost straight, indicating that the nucleus somehow knows where the center of the cell is.

A hint for the underlying mechanism of nuclear centration in fertilized animal zygotes came from careful microscopic observations of some structures growing radially from the nucleus and the nucleus itself moving toward the center. Later, this structure was revealed to be the astral microtubules, which are astral rays of filamentous protein. Given that the nucleus detaches from the cell periphery as the filamentous microtubules grow from the nucleus (Fig. 4.1), pioneering researchers intuitively proposed a hypothetical **pushing model**, in which the growing tip of the microtubules pushes the cell boundary, and the force from the cell boundary pushing back the microtubules moves the nucleus apart from the boundary toward the center of the cell (Chambers 1939) (Fig. 4.2).

Then, the **cytoplasmic pulling model** was proposed (Hamaguchi and Hiramoto 1986) and showed consistency with the initial observations. In this model, the microtubules do not push, but instead are pulled by motor proteins inside the cytoplasm. If we assume a uniform distribution of the motors pulling the microtubules, longer microtubules exhibit stronger pulling forces than shorter ones. When the nucleus is off-center, the microtubules growing toward the center of the cell will be longer, thus moving the nucleus toward the center (Fig. 4.2).

Finally, the most recently proposed **cortex pulling model** is similar to the cytoplasmic pulling model because centration is driven by the force pulling the microtubules (Grill and Hyman 2005 Vallee and Stehman 2005). Instead of assuming force generation throughout the cytoplasm, the cortex pulling model assumes forces generated only at the cell cortex. Likewise, there is a variation in how force is directed toward the center in this model. The most straightforward explanation states that force is proportional to the surface area of the cortex (Grill and Hyman 2005). When the nucleus is off-center, the microtubules growing

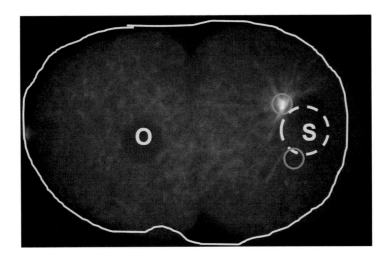

Fig. 4.1 Nuclear centration in the *C. elegans* embryo. Fluorescence microscopy of an embryo expressing green fluorescent protein fused to β-tubulin protein, which is a component of microtubules, highlighted in white. The orange circles indicate the position of the centrosomes, where the lower one is not visible as it is out of frame. The yellow solid line indicates the boundary of the cell. S and O denote the sperm- and oocyte-derived pronuclei, respectively. The boundary of the sperm-derived pronucleus is delimited with the dashed yellow line

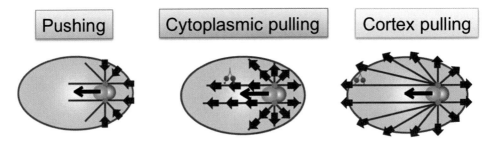

Fig. 4.2 Three models for nuclear centration. The cell structure is colored in green, and the nuclei in blue. The black lines indicate microtubules, and the orange circles represent centrosomes. The red and black arrows indicate the force generation and direction of nuclear migration in the models, respectively

toward the center of the cell cover a larger area of the cortex, and thus the nucleus moves toward the center (Fig. 4.2). For educational purposes, I will adopt a variation of the cortex pulling model that does not consider the surface-area-dependent hypothesis.

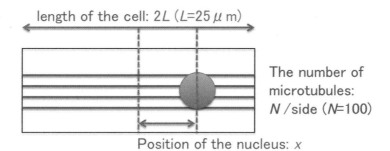

Fig. 4.3 Geometry of one-dimensional model for nuclear centration. The blue circle represents the nucleus, and the green lines represent the microtubules

4.2.2 Constructing One-Dimensional Model for Nuclear Centration

For simplicity, let us construct one-dimensional models for nuclear centration based on the three above-mentioned models. First, we set the length of the cell to be $2L$ and the number of microtubules per side to be N (Fig. 4.3). Then, we will consider what forces are generated when the nucleus is located at position x from the center and how these forces relate to the movement of the nucleus. As solving the former question depends on the model, let us first develop the latter, relating forces to movement.

4.2.2.1 Modeling Forces to Move the Nucleus Using Stokes' Law

The forces acting on the nucleus should be determined to derive the nucleus movement. Based on what we learned in the previous chapter, the Reynolds number of general movements inside the cell is very low, and thus is viscous dominant. In a viscous-dominant regime, a spherical object like the nucleus moves inside a viscous fluid according to the Stokes' law:

$$\mathbf{F} = 6\pi\eta R\mathbf{v}, \tag{4.1}$$

where, $\mathbf{F}$ is the frictional force acting on the object, known as Stokes' drag, R is the radius of the sphere, and $\mathbf{v}$ is the velocity of the object. In other words, drag coefficient k of the sphere is equal to $6\pi\eta R$. In general, the Navier–Stokes equations constitute a physical law that describes the motion of objects in viscous fluids, whereas the Stokes' law is a specific case that neglects the inertia and assumes the incompressibility of the fluid.

Returning to our example, let us model the relationship between the force and velocity of the nucleus by assuming the nucleus is a sphere moving in a viscous medium. With these assumptions, we apply Stokes' law $\mathbf{F} = 6\pi\eta R\mathbf{v}$. As the dynamic viscosity of the cytoplasm of the *C. elegans* embryo is $\eta \approx 1$ [Pa s] (Daniels et al. 2006), and the radius of the nucleus in the embryo is $R \approx 5$ [μm] $= 5 \times 10^{-6}$ [m], $\mathbf{F}$ [N] $\approx \mathbf{v}$ [m/s] $\times 10^{-4}$ [N s/m]. Therefore, the nucleus moves with speed $v = 0.1$ [μm/s] $= 10^{-7}$ [m/s] when applying a force $F = 10^{-11}$ [N] $= 10$ [pN].

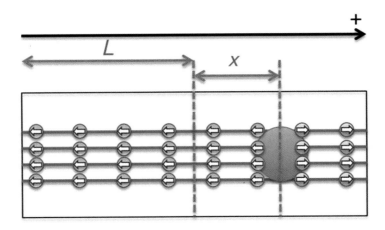

Fig. 4.4 Geometry of cytoplasmic pulling model. The red circles are the force generators, whose force direction is indicated by the inner arrows. The blue circle represents the nucleus, and the green lines represent the microtubules

4.2.2.2 Force Generation in the Cytoplasmic Pulling Model

Let us move on to the calculation of forces generated in each model. I will start from the simplest one, which is the cytoplasmic pulling model. In this model, we consider the force generators located throughout the cytoplasm and pulling the microtubules elongating from the nucleus. Thus, in a simple scenario, the force generated on each microtubule is proportional to its length (Fig. 4.4), and hence we will require two parameters: the force applied by each generator (f_{cyto} [N]) and the density of the force generator per unit length on a microtubule (c [m^{-1}]). When the nucleus (i.e., the ends of all microtubules) is located at position x [m] from the center of the cell, the force to move it ($F_{cyto\text{-}pull}$) can be calculated by subtracting the force toward the minus ($F_{cyto\text{-}minus}$) from that toward the plus direction ($F_{cyto\text{-}plus}$). In addition, the force toward the plus direction is determined by multiplying the force applied by each generator (f_{cyto}) times the density of the generator (c) times the length of each microtubule ($L - x$) and times the number of microtubules (N): $F_{cyto-plus} = f_{cyto}d(L - x)N$. Similarly, $F_{cyto-minus} = f_{cyto}c(L + x)N$. Therefore,

$$F_{cyto-pull} = F_{cyto-plus} - F_{cyto-minus} = f_{cyto}c(L - x)N - f_{cyto}c(L + x)N$$
$$= -2f_{cyto}cxN. \tag{4.2}$$

By substituting Eq. (4.2) into Eq. (4.1), we can obtain the relationship between the position and one-dimensional velocity v of the nucleus as

$$v = -\left(f_{\text{cyto}} cN/3\pi\eta R \right)x. \tag{4.3}$$

4.2.2.3 Force Generation in the Pushing Model

In the pushing model, the force is generated when the microtubule reaches the boundary of the cell and tries to elongate more. The maximum force with which a microtubule can push the cell boundary equals to the maximum force that the microtubule can resist against the compression force. Such force is known as buckling, which is the force required to buckle a rod. For an elastic rod, the buckling force (f_{buckle}) is defined as

$$f_{\text{buckle}} = \pi^2 \kappa / l_r^2, \tag{4.4}$$

where l_r is the length of the rod and κ is the flexural rigidity of the rod (Howard 2001). This formula applies to the case when both ends of the rod are free to pivot. The flexural rigidity of a microtubule (κ_{MT}) is approximately 30×10^{-24} N m^2 (Howard 2001).

For our pushing model, we assume that the force to elongate a microtubule is large enough to freely elongate and push against the cell boundary until the pushing force reaches the buckling force of the microtubule. Similar to the cytoplasmic pulling model, we calculate the force to move the nucleus (F_{push}) by subtracting the force toward the minus ($F_{\text{push}-\text{minus}}$) from that toward the plus direction ($F_{\text{push-plus}}$). Likewise, the force toward the plus direction is determined by multiplying the pushing force per microtubule against the cell boundary at position $-L$ times the number of microtubules reaching the boundary (Fig. 4.5), and thus $F_{\text{push-plus}} = N\pi^2\kappa_{\text{MT}}/(L + x)^2$. Similarly, $F_{\text{push-minus}} = N\pi^2\kappa_{\text{MT}}/(L - x)^2$. Therefore,

$$F_{\text{push}} = F_{\text{push}-\text{plus}} - F_{\text{push}-\text{minus}} = N\pi^2\kappa_{MT} \times \left[1/(L + x)^2 - 1/(L - x)^2 \right]. \tag{4.5}$$

By substituting Eq. (4.5) with Eq. (4.1), we can obtain the relationship between the position and one-dimensional velocity of the nucleus as

$$v = (N\pi\kappa_{\text{MT}}/6\eta R) \times \left[1/(L + x)^2 - 1/(L - x)^2 \right] \tag{4.6}$$

4.2.2.4 Force Generation in the Cortex Pulling Model (an Educational Version)

Finally, let us consider force generation for the cortex pulling model. To provide a pedagogic approach for quantitative biology, I am not modeling the force strictly as proposed in research, but purposely exaggerating the differences in the outcomes among the presented quantitative models.

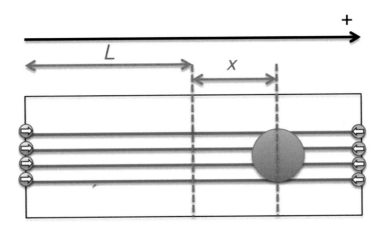

Fig. 4.5 Geometry of the pushing model. The red circles are the force generators, whose force direction is indicated by the inner arrows. The blue circle represents the nucleus, and the green lines represent the microtubules

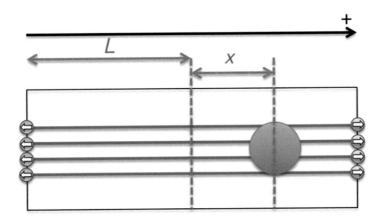

Fig. 4.6 Geometry of the cortex pulling model. The red circles are the force generators, whose force direction is indicated by the inner arrows. The blue circle represents the nucleus, and the green lines represent the microtubules

In this model, the cortex pulls the nucleus by applying a constant force through each microtubule. Notice that for the other models (i.e., cytoplasmic pulling and pushing models) we considered the microtubules to always contact the cortex. If we assume this condition in the cortex pulling model, we will not see the movement of the nucleus, because the force will be proportional to the number of microtubules reaching the cortex. Consequently, the force toward the left and right will be always balanced, regardless of the position of the nucleus (Fig. 4.6). In the conventional version of the cortex pulling model, the pulling force is larger towards the larger space by letting, for instance, the force to be proportional to the area of the cortex covering any space (Grill and Hyman 2005). However, in this textbook, I introduce a variation of the cortex pulling model for educational purposes as follows.

In the proposed version of the cortex pulling model, I consider that the population of longer microtubules is smaller than that of the shorter ones, because a microtubule cannot elongate forever and starts shrinking at a constant frequency (catastrophe rate) (Mitchison and Kirschner 1984). Here, I only consider a constant catastrophe rate F_{cat} [s^{-1}] and that a microtubule encountering catastrophe remains at that length. This assumption follows what is known as the Poisson process. In reality, however, the length decreases after catastrophe and may subsequently re-elongate in vivo. After 1 s, the number of microtubules still elongating will be $(1 - F_{cat})$-fold. Therefore, after t seconds, the number of surviving microtubules will decay exponentially following $(1 - F_{cat})^t$. The time a microtubule continues growing follows exponential distribution

$$P(t) = F_{cat} \exp\left(-F_{cat}\, t\right). \tag{4.7}$$

Hence, the probability for a microtubule to grow for more than t_0 seconds is

$$P(t \geq t_0) = \int_{t_0}^{\infty} F_{cat} \times \exp\left(-F_{cat}t\right)dt = \exp\left(-F_{cat}t_0\right). \tag{4.8}$$

In addition, I set elongation speed V_g [μm/s]. Thus, the probability for a microtubule having length l_{MT} equal or higher than l_0 [μm] is

$$P(l_{MT} \geq l_0) = \exp\left(-F_{cat}\, l_0/V_g\right), \tag{4.9}$$

considering the relationship between time, elongation speed, and length of the microtubule given by $l_{MT} = V_g \times t$. From the total number, N, of microtubules, $N \times \exp(-F_{cat}\, l_0/V_g)$ microtubules reach l_0 micrometers apart from the cortex. Moreover, I consider each microtubule reaching the cortex to be pulled with force f_{cort}. Therefore,

$$F_{cort-plus} = f_{cort}N \times \exp\left[-F_{cat}(L - x)/V_g\right], \tag{4.10}$$

$$F_{cort-minus} = f_{cort}N \times \exp\left[-F_{cat}(L + x)/V_g\right], \tag{4.11}$$

and

$$\begin{aligned}
F_{cort-pull} &= F_{cort-plus} - F_{cort-minus} \\
&= f_{cort}N\left\{ \exp\left[-F_{cat}(L - x)/V_g\right] - \exp\left[-F_{cat}(L + x)/V_g\right]\right\}
\end{aligned} \tag{4.12}$$

By substituting Eq. (4.10) with Eq. (4.1), we will obtain the relationship between the position and one-dimensional velocity of the nucleus as

$$v = \left(f_{\text{cort}} \, N / 6\pi\eta R \right) \times \left\{ \exp\left[-F_{\text{cat}}(L-x)/V_{\text{g}} \right] - \exp\left[-F_{\text{cat}}(L+x)/V_{\text{g}} \right] \right\} \quad (4.13)$$

4.3   Calculating the Movement of the Nucleus

In the previous sections, we derived the relationships between position x and velocity v of the nucleus for each model (Eqs. 4.3, 4.6, and 4.13). As the velocity expresses the rate of change in the position, we can determine the position evolution over time. Specifically, once we set the initial position of the nucleus to be $x = x_0$, we can calculate the velocity at that position as $v_0 = f(x_0)$, which indicates that the velocity is a function of x, as defined in Eqs. (4.3), (4.6), and (4.13). When the nucleus moves with speed v_0 during time Δt, the nucleus is displaced by distance $v\Delta t$, and the position of the nucleus after Δt is $x_{0+\Delta t} = x_0 + v_0\Delta t$. At this new position, the velocity is determined again, and we can move the nucleus further from this position. By repeating this procedure, we can obtain the movement of the nucleus as time passes. It should be noted that this calculation can result in inaccuracy because the position of the nucleus moving during time Δt continuously evolves (because it is moving!), and consequently the velocity also changes. However, if Δt is sufficiently short, the position variation during Δt is small, and the velocity can be considered as constant during this short period. Below, I will explain how we can judge whether Δt is sufficiently short.

4.4   Model Implementation in Microsoft Excel

Now we are prepared to proceed with the actual calculations. Usually, calculations are performed using program codes. As this textbook is intended for beginners in quantitative analysis, I assume you have never written a program before. Instead, I consider you have some experience using a spreadsheet software such as Microsoft Excel. Therefore, to get familiar with model implementation in a computer, let us first use the Microsoft Excel software.

First, create a new document and input the parameters for the three models in the spreadsheet, as listed in Table 4.1. Therefore, in our example, we should input the values from cells D2 to D11, as shown in Fig. 4.7. Note that the values are expressed as multiples of the units to represent meters [m] as micrometers [μm] and newtons [N] as piconewtons [pN], and consequently have values suitable for the scale of a typical cell. In addition, note that μ (micro) stands for multiplier 10^{-6}, and p (pico) for multiplier 10^{-12}.

Table 4.1 Model parameters

Symbol	Description	Unit	Value
General parameters			
L	half-length of the cell	m	25×10^{-6}
N	number of microtubules per side	–	100
R	radius of the nucleus	m	5×10^{-6}
η	dynamic viscosity of the cytoplasm	N s/m^2	1 (Daniels et al. 2006)
dt	timestep for simulation	s	2
Cytoplasmic pulling model			
f_{cyto}	force produced per generator	N	1×10^{-12}
c	density of the force generator on a microtubule	m^{-1}	0.01×10^6
Pushing model			
κ	flexural rigidity of a microtubule	N m^2	30×10^{-24} (Howard 2001)
Cortex pulling model			
f_{cort}	force produced per generator	N	1×10^{-12}
V_g	elongation speed of a microtubule	m/s	1×10^{-6}
F_{cat}	catastrophe rate of a microtubule	s^{-1}	0.2^{a}

^{a}The experimental value is approximately 0.01 s^{-1}

	A	B	C	D	E
1		description	symbol	value	unit
2	model-independent parameters	half length of the cell	L	25	µm
3		number of microtubules per side	N	100	
4		radius of the nucleus	R	5	µm
5		dynamic viscosity of the cytoplasm	η	1	pN s/µm^2
6		unit time of the simulation	dt	2	s
7	cytoplasmic pulling model	force generated by one force generator	fcyto	1	pN
8		density of the force generator on a microtubule	c	0.01	/µm
9	pushing model	flexural rigidity of a microtubule	κ	10	pN µm^2
10	cortical pulling model	force generated by one force generator	fcort	1	pN
11		elongation speed of a microtubule	Vg	1	µm/s
12		catastrophe rate of a microtubule	Fcat	0.2	/s

Fig. 4.7 Excel snapshot of input model parameters

H	I	J	K	L
Cytoplasmic Pulling Model				
time	position	veclocity		
0	20	=−(($D$7*$D$8*$D$3)/(3*PI()*$D$5*$D$4))*I3		

Fig. 4.8 Excel snapshot: velocity calculation for cytoplasmic pulling model

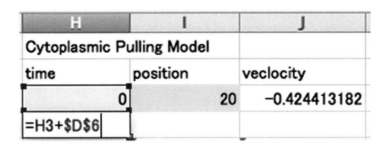

H	I	J
Cytoplasmic Pulling Model		
time	position	veclocity
	0	20 -0.424413182
=H3+$D$6		

Fig. 4.9 Excel snapshot: time increment for cytoplasmic pulling model

4.4.1   Implementation of Cytoplasmic Pulling Model

We obtained the relationship between position x and velocity v of the nucleus for the cytoplasmic pulling model in Eq. (4.3). Hence, we can calculate the velocity from any position of the nucleus. Now, let us calculate the velocity and position in columns H to J, as shown in Fig. 4.8. First, input the values for initial time 0 s and position 20 μm at cells H3 and I3, respectively. Note that the initial position can be anywhere you like within the cell boundary. To calculate the velocity at this position, write

```
=-(($D$7*$D$8*$D$3)/(3*PI()*$D$5*$D$4))*I3
```

in cell J3 (Fig. 4.8). Note that this formula is the same as Eq. (4.3) because the values of f_{cyto}, c, N, η, R, and x are placed in cells D7, D8, D3, D5, D4, and I3, respectively (you can verify this by seeing Fig. 4.7). In addition, PI() is the function that retrieves constant π in Excel.

Next, let us move the nucleus. As we now know the velocity of the nucleus at the initial time $(v_{t=0})$, we can determine the nucleus displacement at time Δt as distance $v_{t=0} \times \Delta t$. As an initial attempt, set $\Delta t = 2$ s (indicated as dt in cell D6 of Fig. 4.7). Therefore, the next time value will be $0 + \Delta t = 2$, which can be expressed in cell H4 as (Fig. 4.9):

```
=H3+$D$6
```

The corresponding position at 2 s is $x_{t=0} + v_{t=0} \times \Delta t$, which can be calculated in cell I4 by writing (Fig. 4.10):

```
=I3+J3*$D$6
```

Now that we have an updated position of the nucleus, we can update its velocity, which depends on the position. Therefore, input a definition in cell J4 that is analogous to the definition in J3. This time, you do not have to type, but just copy the contents of cell J3 into J4 (Fig. 4.11). In Excel, if you copy the contents of a cell into another, the related cells in

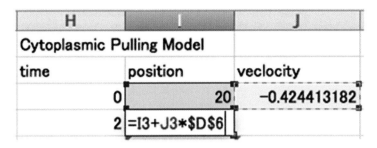

H	I	J
Cytoplasmic Pulling Model		
time	position	veclocity
0	20	−0.424413182
2	=I3+J3*$D$6	

Fig. 4.10 Excel snapshot: position calculation for cytoplasmic pulling model

H	I	J	K	L
Cytoplasmic Pulling Model				
time	position	veclocity		
0	20	−0.424413182		
2	19.15117364	=−(($D$7*$D$8*$D$3)/(3*PI()*$D$5*$D$4))*I4		

Fig. 4.11 Excel snapshot: copied velocity calculation for cytoplasmic pulling model

the formulas are updated in a process called relative cell reference. In this case, cell I3 is referred in J3, and hence cell I4 is automatically referred when the contents are copied into cell J4. This feature is useful in this case because we want to use the current position (cell I4) to update the velocity. However, this feature is not always useful. For instance, the reference to general parameters, e.g., f_{cyto} and c, should remain fixed to the cells containing them, e.g., D7 and D8, and it would be wrong if these cells' references change to D8 and D9. To prevent relative cell reference in these cases, place symbol "$" in front of both the column letter and row number of the cell whose reference should remain fixed, e.g., $D$7 and $D$8.

We can keep on updating the position of the nucleus by copying the contents of cells H4, I4, and J4 into the subsequent rows, for example, the next 120 rows. Then, we can draw a graph by plotting the position of the nucleus against time to visualize the behavior of the numerical model (Fig. 4.12). From the graph, we see that the position eventually reaches 0, indicating that the nucleus is located at the center of the cell. In addition, the nucleus reaches the center in approximately 200 s. Now, we have created our first simulation in this textbook. I hope the process above is easy for you to follow, even if you have no previous experience with numerical simulations. The main point in this example is to highlight that simulations are not difficult to develop and implement. Even a familiar software such as Microsoft Excel can be used as a simulator!

Now, it is worth discussing the value of Δt (cell D6 in Fig. 4.7). I previously mentioned that this value should be sufficiently small. Let us evaluate what happens if we change that value using our simulation program. It is easy for us to do this because we can change this

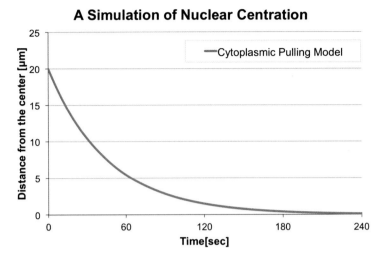

Fig. 4.12 Excel plot: nuclear centration using the cytoplasmic pulling model

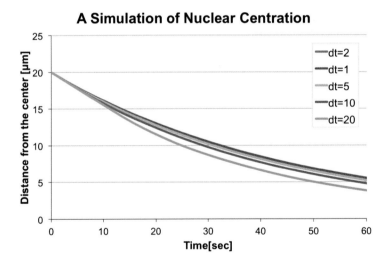

Fig. 4.13 Excel plot: comparison among different Δt values

value at cell D6. Figure 4.13 shows the simulation results for $\Delta t = 1, 2, 5, 10,$ and 20 s. The results for $\Delta t = 1$ and 2 s are indistinguishable, and we can only see the result for $\Delta t = 1$ s because it overlays that for $\Delta t = 2$ s. This suggests that $\Delta t = 2$ is sufficiently small, because further decreasing the value does not affect the results. In contrast, larger values of Δt retrieve different results from those for $\Delta t = 1$ and 2 s. Hence, any Δt above 5 s is inappropriately large, and the results do not accurately reflect the model behavior. The shorter the timestep for simulation, the more accurate the results. However, shortening the

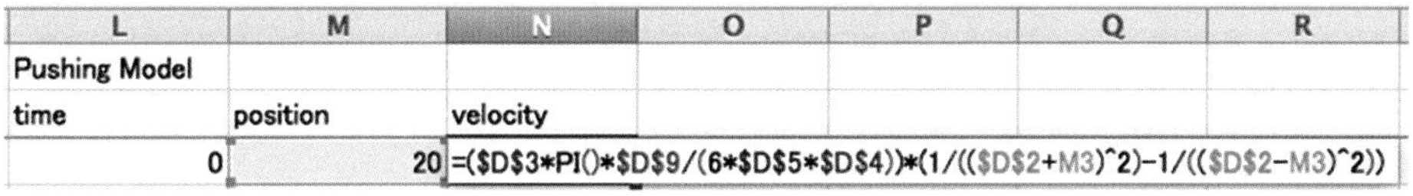

L	M	N	O	P	Q	R
Pushing Model						
time	position	velocity				
0	20	=($D$3*PI()*$D$9/(6*$D$5*$D$4))*(1/(($D$2+M3)^2)-1/(($D$2-M3)^2))				

Fig. 4.14 Excel snapshot: velocity calculation for pushing model

timestep implies more computational cost (time) to calculate the results for a given period, as you can realize by the increasing number of rows required to calculate the results for a period (e.g., 100 s) as the timestep shortens. Hence, when conducting this type of simulation, it is important that you consider the tradeoff between timestep Δt and the computational cost in terms of data and runtime.

4.4.2 Implementation of Pushing Model

Similar to the cytoplasmic pulling model, let us model the pushing model in the same spreadsheet from columns L to N. The relationship between position x and velocity v of the nucleus in the pushing model is given by Eq. (4.6).

As we did for the cytoplasmic pulling model, we input initial time 0 s and position 20 μm at cells L3 and M3, respectively. To calculate the velocity at this position, write in cell N3 (Fig. 4.14):

```
=($D$3*PI()*$D$9/(6*$D$5*$D$4))*(1/($D$2+M3)^2-1/($D$2-M3)^2)
```

Here, "^2" represents the squaring of the preceding term.

After calculating the velocity, we calculate the movement of the nucleus as we did for the cytoplasmic pulling model. Briefly, we input in cell L4 the next time value as (Fig. 4.15):

```
=L3+$D$6
```

and update the position at that time in cell M4 as (Fig. 4.16):

```
=M3+N3*$D$6
```

For the remaining of the cells in columns L, M, and N, just copy and paste the contents of cells L4, M4, and N3, respectively.

Now, let us see how the pushing model behaves by plotting the position of the nucleus against time (Fig. 4.17). We can see that the nucleus also moves to the center using this model, but it reaches faster than when using the cytoplasmic pulling model, after approximately 150 s.

L	M	N
Pushing Model		
time	position	velocity
0	20	−4.13707675
=L3+$D$6		

Fig. 4.15 Excel snapshot: time increment for pushing model

L	M	N
Pushing Model		
time	position	velocity
0	20	−4.13707675
2	=M3+N3*$D$6	

Fig. 4.16 Excel snapshot: position calculation for pushing model

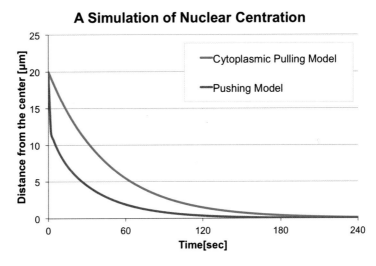

Fig. 4.17 Excel plot: nuclear centration using the pushing model (red) and cytoplasmic pulling model (blue)

P	Q	R	S	T	U	V	W	X	Y	Z	AA
Cortex Pulling Model											
time	position	velocity									
0	20	=($D$10*$D$3/(6*PI()*$D$5*$D$4))*(EXP(−$D$12*($D$2−Q3)/$D$11)−EXP(−$D$12*($D$2+Q3)/$D$11))									

Fig. 4.18 Excel snapshot: velocity calculation for cortex pulling model

4.4.3 Implementation of Cortex Pulling Model

Finally, let us model the cortex pulling model in the same spreadsheet by using columns P to R. The relationship between position x and velocity v of the nucleus in the cortex pulling model is given by Eq. (4.13). Start by inputting initial time 0 s and position 20 μm at cells P3 and Q3, respectively. To calculate the velocity at this position, write in cell R3 (Fig. 4.18):

```
=($D$10*$D$3/(6*PI()*$D$5*$D$4))*(EXP(-$D$12*($D$2-Q3)/$D$11)-EXP(-$D$12*
($D$2+Q3)/$D$11))
```

Here, "EXP" is the exponential function with the natural logarithm as base.

After calculating the velocity, we calculate the movement of the nucleus as we did for the other models. We update the time and position in cells P4 and Q4, respectively, as

```
=P3+$D$6, and
=Q3+R3*$D$6
```

For the remaining cells in columns P, Q, and R, copy and paste the contents of cells P4, Q4, and R3, respectively.

Here, we encounter a problem. Values are not computed after 20 s, and message "#NUM!," which indicates a very large number, appears instead (Fig. 4.19). It is unrealistic that position and velocity become very large in the model, but this is caused by the position of the nucleus, which surpasses 25 μm after 10 s, indicating that the nucleus crossed the cell boundary.

Until now, we have not considered the case when the nucleus crosses the cell boundary. When the nucleus crosses the right boundary (i.e., $x > 25$ μm), the right section of the cortex should not pull the nucleus anymore (i.e., $F_{\text{cort-plus}} = 0$). To introduce this constraint, we should use conditional branching. Specifically, we can use function "IF" in Excel. Let us use column S to calculate $F_{\text{cort-plus}}$. In cell S3, write (Fig. 4.20):

```
=IF(Q3>$D$2,0,$D$10*$D$3*(EXP(-$D$12*($D$2-Q3)/$D$11)))
```

P	Q	R
Cortex Pulling Model		
time	position	velocity
0	20	0.390201268
2	20.78040254	0.456155227
4	21.69271299	0.54750431
6	22.78772161	0.681592874
8	24.15090736	0.895261546
10	25.94143045	1.280816614
12	28.50306368	2.137943685
14	32.77895105	5.028044324
16	42.83503969	37.57168924
18	117.9784182	126395514.8
20	252791147.5	#NUM!
22	#NUM!	#NUM!
24	#NUM!	#NUM!
26	#NUM!	#NUM!

Fig. 4.19 Excel snapshot: cortex pulling model without considering the cell boundary

P	Q	R	S	T	U	V	W
Cortex Pulling Model							
time	position	velocity	Fcort–plus	Fcort–minus			
0.0	20.0		=IF(Q3>$D$2,0,$D$10*$D$3*(EXP(−$D$12*($D$2−Q3)/$D$11)))				

Fig. 4.20 Excel snapshot: introducing conditional branching for considering the cell boundary

Function "IF" requires three parameters: IF(a,b,c). The first parameter, a, defines the condition; the second one, b, defines the calculation to be conducted if the condition is satisfied; and the third one, c, defines the calculation to be conducted if the condition is not satisfied.

Similarly, for the left boundary, we can use conditional branching as follows: if $x < -25$, then $F_{cort-minus} = 0$, otherwise $F_{cort-minus} = -f_{cort}N \times \exp[-F_{cat}(L + x)/V_g]$ (see Eq. 4.11), and introduce it in cell T4:

```
=IF(Q3<-$D$2,0,$D$10*$D$3*(EXP(-$D$12*($D$2+Q3)/$D$11)))
```

P	Q	R	S	T
Cortex Pulling Model				
time	position	velocity	Fcort–plus	Fcort–minus
0	20.00	0.39	36.79	0.01
2	20.78	0.46	43.00	0.01
4	21.69	0.55	51.61	0.01
6	22.79	0.68	64.25	0.01
8	24.15	0.90	84.38	0.01
10	25.94	0.00	0.00	0.00
12	25.94	0.00	0.00	0.00
14	25.94	0.00	0.00	0.00
16	25.94	0.00	0.00	0.00
18	25.94	0.00	0.00	0.00
20	25.94	0.00	0.00	0.00
22	25.94	0.00	0.00	0.00
24	25.94	0.00	0.00	0.00
26	25.94	0.00	0.00	0.00
28	25.94	0.00	0.00	0.00
30	25.94	0.00	0.00	0.00

Fig. 4.21 Excel snapshot: cortex pulling model after introducing conditional branching to consider the cell boundary

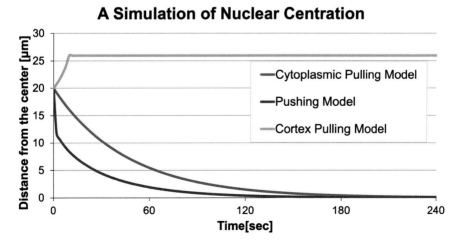

Fig. 4.22 Excel plot: nuclear centration using the pushing model (red), cytoplasmic pulling model (blue), and cortex pulling model (green)

With these conditions, we calculate the velocity according to $v = F_{\text{cort-pull}}/6\pi\eta R = (F_{\text{cort-plus}} - F_{\text{cort-minus}})/6\pi\eta R$ (see Eqs. 4.10–4.13).

This is included in Excel by writing in cell R3:

```
=(S3-T3)/(6*PI()*$D$5*$D$4)
```

With these modifications, the nucleus mostly stays within the boundary (Fig. 4.21).

You might have already noticed that the nucleus does not move to the center using the cortex pulling model (Fig. 4.22). As I mentioned previously, the cortex pulling model introduced here is not as that proposed in research. Instead, I introduced a different way of modeling the forces generated at the cortex to exaggerate the difference in the outcomes among quantitative models, and present a case where the model does not behave as it is supposed to.

In this chapter, I introduced several hypotheses for nuclear centration and formulated some simple models accordingly. Once you formulate your hypotheses, you can conduct a simulation using a familiar software like Microsoft Excel. However, as the hypotheses and their formulation become more sophisticated, it is difficult to conduct a simulation using such type of software. As the next step, I will implement simulations identical to those shown in this chapter, but using a comprehensive programing language called Python. By using Python, you will be able to conduct more complicated simulations.

Questions
1. Change the value of Δt for both the pushing model and cortex pulling model and see how the outcomes are affected. Are the appropriate timesteps, Δt, similar between these models?
2. Change the other parameters in the models and analyze how they affect the outcomes. Before conducting the simulations, predict some plausible outcomes.

Answers
1–2 For both questions, please conduct the simulations to find out the answers.

> **Take-Home Message**
> - My goal was for anyone to be able to experience the derivation and implementation of a numerical model using the familiar Microsoft Excel software. If you come up with a hypothesis on the mechanism underlying any phenomenon of your interest, what you need is to formulate the mechanism using mathematical expressions.
> - I introduced several hypotheses for nuclear centration and formulated them as simply as possible. These examples will be used in the following chapters.

References

Chambers EL. The movement of the egg nucleus in relation to the sperm aster in the echinoderm egg. J Exp Biol. 1939;16:409–24.

Daniels BR, Masi BC, Wirtz D. Probing single-cell microme-chanics in vivo: the microrheology of C. elegans developing embryos. Biophys J. 2006;90:4712–9.

Grill SW, Hyman AA. Spindle positioning by cortical pulling forces. Dev Cell. 2005;8:461–5.

Hamaguchi MS, Hiramoto Y. Analysis of the role of astral rays in pronuclear migration in sand dollar eggs by the colcemid-UV method. Dev Growth Differ. 1986;28:143–56.

Howard J. Mechanics of motor proteins and the cytoskeleton. Sinauer Associates, Inc; 2001.

Mitchison T, Kirschner M. Dynamic instability of microtubule growth. Nature. 1984;312:237–42.

Vallee RB, Stehman SA. How dynein helps the cell find its center: a servomechanical model. Trends Cell Biol. 2005;15:288–94.

Wilson EB. The cell in development and heredity. 3rd ed. The Macmillan Company; 1925.

Implementing Toy Models in Python

5

Contents

> **What You Will Learn in This Chapter**
> Quantitative biology requires computational methods for many tasks, including running simulations, processing images, and conducting statistical analyses. Therefore, computer programming is an essential tool for quantitative biology. In this textbook, I will focus on the Python programming language. This chapter provides a step-by-step introduction for simple simulation using Python based on the same models we implemented in Microsoft Excel in the previous chapter. Thus, you should know the details of the model by now. Constructing the same models with Python will allow readers to take the first steps for programming in Python, a tool that enables a broad range of analyses in quantitative biology.

Learning Objectives

After completing this chapter, readers should be able to

1. Implement the numerical relationships with Python and evaluate the hypothesis.
2. Use `for` loops and branching (`if` statement) to write simple codes in Python.

© Springer Nature Singapore Pte Ltd. 2022   51
A. Kimura, *Quantitative Biology*, Learning Materials in Biosciences,
https://doi.org/10.1007/978-981-16-5018-5_5

Important Concepts Discussed in This Chapter

- *Loops*; These are used to repeat procedures in computer programing, which is important in conducting simulations.
- *Branching*; This is used to change procedures in computer programing, which is another important aspect in conducting simulations.

5.1   Why Do We Need to Learn Programming?

Quantitative biology requires computers for many tasks, including running simulations, processing images, and conducting statistical analyses. A myriad of specialized computer software is available for every type of computation. Therefore, we could select the appropriate software depending on the computation we aim to perform. However, searching and choosing an appropriate software is often time-consuming, and most software platforms are paid. Learning how to use specialized software can also be time-consuming and expensive. In research, we often require linking multiple analyses, for example, taking the outcomes from image processing to run a simulation. In this type of situation, it is more convenient to know how to create software suitable for diverse purposes. By learning programming, you have to neither acquire software, because you create your own programs, nor learn software operation because you created the programs and fully know them. Moreover, you can relate any analyses you need. Of course, creating an analysis tool by yourself is also time-consuming, but it is totally worthwhile, as you will know every detail of the procedures and be able to customize the software to meet all your requirements.

5.2   Why Python?

There are many programming languages, such as C, C++, FORTRAN, MATLAB, and Python. When I started learning programming, I used the C language, which is one of the most basic programming languages, and you can use it for free in most environments. Starting with the C language was a good decision that time, because it requires to describe almost every detail of the implementation. For example, consider solving a quadratic equation of the form $3x^2 - 2bx - 4 = 0$. In C, you need to define every variable of the equation in the program and conduct case analyses based on the sign of the discriminant. You must specify almost all the details to obtain the solution, and thus you will completely know them.

Then, I started using MATLAB. Although it is not free, it notably simplifies the implementation of numerical analyses. Taking the same example of solving the quadratic equation, all you need to do is define the equation as p = [3 -2 4]; and solve it by typing r = roots(p) in MATLAB. Several optional toolboxes, which may cost additional money, are specialized for applications such as statistical analyses and image processing. As MATLAB has many useful functions and tools, I use it in my regular research practice.

On the other hand, Python is one of the most popular programing languages nowadays. It allows to conduct similar analyses as MATLAB for free. Returning to the example of solving the quadratic equation, you can obtain the result by using the solve function after importing the "sympy" library. As Python has a large number of users, one can benefit from collaborative development of many libraries and applications and the availability of numerous forums and tutorials. Therefore, Python is a very good language to start learning programming with.

5.3   Getting Started with Python

Throughout this textbook, we will use the Anaconda distribution of Python 3.x. Anaconda provides a package of important libraries for scientific calculations (e.g., NumPy, SciPy, matplotlib, and IPython) in addition to the base Python distribution. Anaconda can be installed on Windows, macOS, and Linux operating systems.

The installation of Anaconda is straightforward. You should visit www.anaconda. com/download/ from your web browser, choose the target operating system (Windows, macOS, or Linux), click the Download button, and execute the download-guided installation. The latest version should be fine, which is Python 3.6 as I write this chapter. The details of the installation may change over time, but it is always meant to be intuitive. After installation, you should be able to launch the Spyder application, which is an integrated development environment for Python programming.

A few customizations of Spyder are recommended (Kinder and Nelson 2018). First, you should set the working directory. The files you create with Spyder will be saved to this directory by default for easy access to your code source files. This directory can be set from the left panel of the Preferences menu by selecting "Global working directory" and specifying the directory you like. Second, you should enable the interactive graphics, as in some cases, the output of your programs will be graphical items (e.g., plots and images), and making them interactive enables to zoom, move, and rotate them as you wish. To do this, choose "IPython console" from the left panel of the Preferences menu, click the "Graphics" button, and set the "Graphics back end" section to "Automatic." After setting these changes, click "Apply" at the bottom of the panel and restart Spyder to see these changes.

5.4   A Code to Simulate Nuclear Centration

The following is a code to simulate nuclear centration.

Code 5.1 Centration1D.py

```
"""
Centration1D.py
Created on Sat Sep 10, 2016
@author: akkimura (Python 3.5)
Description: simple 1D simulation for nuclear centration
"""
import numpy as np
from numpy import pi, exp, log
import matplotlib.pyplot as plt


#%% Initialize parameters
L = 25 # half length of the cell [um]
N = 100 # number of microtubules (half)
R = 5 # the Stokes radius of the nucleus [um]
eta = 1 # the viscosity of the cytoplasm [pN s/um^2]
f = 1 # pulling force of a single motor [pN]
c = 0.01 # the density of the motors on microtubules [/um]
Vg = 1 # growth speed of microtubule [um/s]
Fcat = 0.2 # the catastrophe frequence of microtubules [/s]
kappa = 10 # flexural rigidity of microtubules [pN um^2]
K = 6*pi*eta * R # coefficient for nucleus drag [pN s/um]
dt = 2 # sec per simulation step [s]
totSTEP = 120 # the number of simulation steps
X0 = 20 # initial position of the nucleus [um]


#%% Initialize variables
Xcyto = np.zeros(totSTEP+1)
Xcort = np.zeros(totSTEP+1)
Xpush = np.zeros(totSTEP+1)
Xcyto[0] = X0
Xcort[0] = X0
Xpush[0] = X0


#%% Calculate
for st in range(totSTEP): # i starts from 0 to totSTEP-1
  # cytoplasmic pulling model
  Fcyto = -2 * f * c * N * Xcyto[st]
  Xcyto[st+1] = Xcyto[st] + dt * Fcyto/K
  # cortical pulling model
  if Xcort[st] > L:
    Fcort_plus = 0
  else:
    Fcort_plus = f * N * (exp(-Fcat*(L-Xcort[st])/Vg))
```

```
if Xcort[st] < -L:
    Fcort_minus = 0
else:
    Fcort_minus = -f * N * (exp(-Fcat*(L+Xcort[st])/Vg))
Fcort = Fcort_plus + Fcort_minus
Xcort[st+1] = Xcort[st] + dt * Fcort/K
# pushing model
Fpush = N * pi * pi * kappa * (1/((L+Xpush[st])**2) - 1/((L-Xpush[st])**2))
Xpush[st+1] = Xpush[st] + dt * Fpush/K


#%% plot results
t_values = np.linspace(0,totSTEP*dt,totSTEP+1)
plt.plot(t_values, Xcyto, color="b", label="Cyto Pull")
plt.plot(t_values, Xcort, color="r", label="Cort Pull")
plt.plot(t_values, Xpush, color="g", label="Push")


# graph modifications
ax = plt.gca() # get current axis
ax.set_title("centration simulation", size=24, weight='bold')
ax.set_xlabel("time [s]", size=18)
ax.set_ylabel("Position of the nucleus [um]", size=18)
plt.legend(loc='center right') # show legends


#save the figure
fig = plt.gcf() # get current figure
plt.savefig("centration_test.png")
```

To run this Python code, you need to either create a new file by choosing "New File" from the File menu of Spyder and type the code or download it to the working directory and open the file by choosing "Open" from the File menu. Once the code appears on the Editor window and is saved, you can run it by choosing "Run" from the Run menu of Spyder, or type `runfile('Centration1D.py')` in the IPython console. Running this code will retrieve a result as that shown in Fig. 5.1.

The result is basically the same as that obtained by using Excel (Fig. 4.22), indicating the consistency of the calculations in Excel and Python.

Let us analyze the code part by part to understand it.

```
"""
Centration1D.py
Created on Sat Sep 10, 2016
@author: akkimura (Python 3.5)
Description: simple 1D simulation for nuclear centration
"""
```

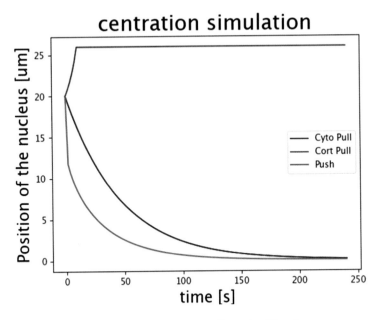

Fig. 5.1 Result of executing nuclear centration code "Centration1D.py"

The first part comprises comments for yourself or colleagues, to record information such as the developer, version, and date of coding. When you enclose text between two pairs of triple quotation marks (''' or " " "), the text will not be executed by Python. This may not be technically accurate, but it is correct in practice. Hence, you can write any relevant information for others to see within your source code.

```
import numpy as np
from numpy import pi, exp, log
import matplotlib.pyplot as plt
```

This second part declares the libraries that you will use in your code. Libraries are like toolboxes, and Python has many of them. Thus, it would be inefficient to have all the toolboxes within reach for your code. Therefore, you should specify every toolbox you will use in the code. Library "numpy" is a library for numerical computations. For example, function "numpy.zeros(N)" from this library creates an array of N components. As typing "numpy" for each of its functions can be laborious, you can abbreviate "numpy" as "np," as defined in the first line. Library "numpy" includes definitions of the ratio of circumference of a circle (π), logarithm, and exponential functions. You can use "numpy.pi" (or "np.pi") for constant π, but this is also laborious as many instances may occur throughout the code. Hence, the second line of this part imports the definitions of π, logarithm, and exponential functions to be used directly as "pi," "exp," and "log," respectively, without placing "numpy." or "np." in front of them. The third line allows to use other library, "matplotlib. pyplot," as a tool to plot graphs, and we abbreviate it as "plt."

```
#%% Initialize parameters
L = 25 # half length of the cell [um]
N = 100 # number of microtubules (half)
R = 5 # the Stokes radius of the nucleus [um]
eta = 1 # the viscosity of the cytoplasm [pN s/um^2]
f = 1 # pulling force of a single motor [pN]
c = 0.01 # the density of the motors on microtubules [/um]
Vg = 1 # growth speed of microtubule [um/s]
Fcat = 0.2 # the catastrophe frequence of microtubules [/s]
kappa = 10 # flexural rigidity of microtubules [pN um^2]
K = 6*pi*eta * R # coefficient for nucleus drag [pN s/um]
dt = 2 # sec per simulation step [s]
totSTEP = 120 # the number of simulation steps
X0 = 1 # initial position of the nucleus [um]
```

This part sets the parameters for the nuclear centration simulation. The names and values of the parameters are the same as those we used in Chap. 4 (Table 4.1), where we conducted the simulation in Excel. Here, the hash (#) character indicates the beginning of a comment until the end of line, and hence the text after it and until a new line will not be considered for execution. Note how comments are important for you and others to understand the code.

```
#%% Initialize variables
Xcyto = np.zeros(totSTEP+1)
Xcort = np.zeros(totSTEP+1)
Xpush = np.zeros(totSTEP+1)
Xcyto[0] = X0
Xcort[0] = X0
Xpush[0] = X0
```

This part creates arrays to store the results. Arrays are like the groups of cells in Excel, where we put the results in the previous chapter. Array Xcyto will contain the results of the cytoplasmic pulling model, and zeros is a function of "numpy" to create a matrix of zeros of any size. Hence, Xcyto = np.zeros(totSTEP+1) means that Xcyto is an array with 121 zeros (i.e., [0, 0, 0, ..., 0]), because totSTEP is 120, as defined in the previous part. Now you may realize an advantage of using Python over Excel. In Excel, you should copy cells to prepare them for placing the results, but in Python, all you have to do is change the value of totSTEP.

Xcyto[i] is the i-th component of the array, which starts from index 0 in Python. Therefore, "Xcyto[0]" is the first component of the array. As X0 is the initial position of the nucleus in the three models, we assign X0 to Xcyto[0], Xcort[0], and Xpush[0].

```
#%% Calculate
for st in range(totSTEP): # i starts from 0 to totSTEP-1
  # cytoplasmic pulling model
  Fcyto = -2 * f * c * N * Xcyto[st]
  Xcyto[st+1] = Xcyto[st] + dt * Fcyto/K
  # cortical pulling model
  if Xcort[st] > L:
    Fcort_plus = 0
  else:
    Fcort_plus = f * N * (exp(-Fcat*(L-Xcort[st]))/ Vg))
  if Xcort[st] < -L:
    Fcort_minus =0
  else:
    Fcort_minus = -f * N * (exp(-Fcat*(L+Xcort[st])/ Vg))
  Fcort = Fcort_plus + Fcort_minus
  Xcort[st+1] = Xcort[st] + dt * Fcort/K
  # pushing model
  Fpush = N * pi * pi * kappa * (1/((L+Xpush[st])**2) - 1/((L-Xpush[st])**2))
  Xpush[st+1] = Xpush[st] + dt * Fpush/K
```

This is the main part of the code, as it contains the necessary calculations for the simulation. We can code the calculations for the three models in a single code structure. Loop for is one of the most important commands in any program, as it repeats procedures described in the subsequent indented lines. The description of for i in *array* means that the procedures will be repeated while the change in variable i is defined in *array*. For instance, function range creates an array as an integer interval, and in our case, range (totSTEP) retrieves array [0, 1, 2, ..., 119]. Hence, for st in range(totSTEP) indicates that the procedure will be repeated 120 times as *st* changes from 0 to 119 at each iteration.

Consider the lines after for. The subsequent three lines (including the comment line) aim to calculate the nuclear position for the cytoplasmic pulling model. First, the force to move the nucleus when it is at position "Xcyto[*st*]" is calculated. At the first iteration, $st = 0$ and Xcyto[*st*] = X0. In the next line, the position of the nucleus at the (st + 1)-th step is calculated from the force and current position of the nucleus. This new position will be stored in the (st + 1)-th component of array "Xcyto," just as we put numbers in the cells when conducted the simulation in Excel. Here, you will notice another advantage of Python over Excel: defining formulas is much easier and more straightforward in Python.

Another important command in any program is if, which is used for conditional branching. In the cortical pulling model, we need to set the pulling force from the right (plus) side zero (i.e., Fcort_plus = 0) if the nucleus exceeds the right boundary (i.e., Xcort[st] > L) and similar for the left (minus) side. We did this in our simulation using Excel (see Sect. 4.4.3). This process, where we want the program to function in a condition-dependent manner, is called conditional branching and is realized with `if` (condition): `else:` with Python.

```
#%% plot results
t_values = np.linspace(0,totSTEP*dt,totSTEP+1)
plt.plot(t_values, Xcyto, color="b", label="Cyto Pull")
plt.plot(t_values, Xcort, color="r", label="Cort Pull")
plt.plot(t_values, Xpush, color="g", label="Push")


# graph modifications
ax = plt.gca() # get current axis
ax.set_title("centration simulation", size=24, weight='bold')
ax.set_xlabel("time [s]", size=18)
ax.set_ylabel("Position of the nucleus [um]", size=18)
plt.legend(loc='center right') # show legends


#save the figure
fig = plt.gcf() # get current figure
plt.savefig("centration_test.png")
```

After the calculations, this last part of the code aims to plot the results as a graph. This is also a benefit of coding a program over using a spreadsheet software, as graphs are created automatically after you code them. Command `plt.plot` (which is the abbreviation for of `matplotlib.pyplot.plot`) draws a scatter plot with the horizontal axis specified by the first argument (e.g., `t_values` in the third line), and the vertical axis specified by the second argument (e.g., `Xcyto` in the third line). Array `t_values` defines the simulation timesteps with `totSTEP + 1` components for time starting from 0 (0 s) until `totSTEP*dt` (240 s) at simulation interval `dt` (2 s). This array is defined by using function `np.linspace`. Upon plotting graphs, we can specify visual characteristics such as colors of the curves, data labels, and many other aspects, using different options. Similarly, we can specify the title, *x*- and *y*-axis labels, and legends as shown in lines 7–12 of this part (the lines under comment "`# graph modifications`"). Finally, you can save the graph as an image file, e.g., "centration_test.png," as shown in the last line of the code.

You should now completely understand what is written in Code 5.1 and be able to create similar simulation codes on your own. Still, to suit the code for your purposes or creating your own codes, you may require additional knowledge about Python. Here, I explained only the basics of Python through this illustrative code, whose many details were omitted. But you do not have to worry. Once you clearly state your problem and have a basic framework for creating codes, which I expect you acquire from this textbook, you can find the tools (e.g., commands, functions, and libraries) easily by searching in the Internet or other sources. In fact, many people share their experience and tips on Python in the Internet. Therefore, simply searching for applications you need, such as typing "Python plot graph" on a search engine, will provide you various examples on related codes that can be applied to yours.

Questions
1. Change the value of Δt and other parameters in the models and analyze how they affect the outcome. Implement and compare the same procedures in Excel and Python and discuss the benefits of each approach.

Answers
1. Please conduct the simulation yourself and find the answer.

Take-Home Message
- Python is one of the most popular programing languages these days. Learning programing with Python will enable you to conduct various analyses, including simulation, image processing, and statistical analyses.
- The code presented in this chapter replicates the calculation using a spreadsheet software, Microsoft Excel, in Chap. 4. By comparing the two procedures, you will understand every detail of this simple code.
- Once you understand and write a simple code, it will be easy for you to arrange the code and create more sophisticated codes. While the detailed instructions of Python are not provided in this textbook, you can find many materials available on the Internet.

Reference

Kinder JM, Nelson P. A student's guide to python for physical modeling. Princeton University Press; 2018.

Differential Equations to Describe Temporal Changes

6

Contents

What You Will Learn in This Chapter

As biological systems are dynamic, modeling temporal changes is the key in quantitative biology. Modeling of temporal changes refers to the prediction of the future. Differential equation is a powerful mathematical tool used for this purpose, and thus it is considered a fundamental tool in quantitative biology. In this chapter, elementary knowledge on differential equations is reviewed. This includes a linear stability analysis, which has the ability to convey the behavior of mathematical models, based on differential equations. We will also use Python to solve, or simulate the results of differential equations.

© Springer Nature Singapore Pte Ltd. 2022
A. Kimura, *Quantitative Biology*, Learning Materials in Biosciences,
https://doi.org/10.1007/978-981-16-5018-5_6

Learning Objectives

After completing this chapter, readers should be able to

1. Obtain information on the equilibrium points and the stability of the points from a simple differential equation.
2. Solve a simple differential equation analytically.
3. Calculate the consequences of differential equations with both Euler and Range–Kutta methods using Python.

Important Concepts Discussed in This Chapter

- *Differential equation*; equation that relates functions and their derivatives.
- *Equilibrium point*; the point where the derivatives of a differential equation are zero. If the system reaches the equilibrium point, the variables do not change. There are stable and unstable equilibrium points.
- *Euler and Runge–Kutta methods*; methods to numerically solve differential equations.

6.1   Why the Use of a Differential Equation?

6.1.1   What Is a Differential Equation?

A differential equation is a class of mathematical equations including functions and their derivatives. The derivative of a function means the rate of change of the function against a change in its argument. When time is an argument of the function, the differential equation defines a derivative (the rate of change over time), and thus can predict a future from the current status. Therefore, differential equation is useful for modeling dynamic systems, such as biological systems.

For example, suppose we want to predict a value, $x(t)$, which changes over time, t (e.g. the expression level of gene X). Let us consider a simple situation in which this change depends only on the current value of x, and thus can be defined as a function of x, $F(x)$. This function $F(x)$ defines the rate of change of x. Supposing the current time is $t = t_0$, and x is $x(t_0)$, what will be the value of x at Δt later? Let us represent the change of $x(t)$ as Δx. $\Delta x = x(t + \Delta t) - x(t)$. Because $F(x)$ is the rate of change, $\Delta x/\Delta t \fallingdotseq F(x)$, when Δt is small. The derivative is defined as $\frac{dx}{dt} = \lim_{\Delta t \to 0} \frac{\Delta x}{\Delta t}$. From the definition of $F(x)$, $\frac{dx}{dt} = F(x)$. This equation is a differential equation as it relates a function to its derivative.

Consider a simple example:

$$dx/dt = F(x) = 25 - x. \tag{6.1}$$

If the current value of x is less than 25, for example, if $x = 10$, $dx/dt = F(x) = 25-10 = +15$; because dx/dt is a positive value in this case, x will increase as time progresses (t increases). On the other hand, if current x is greater than 25, for example, if $x = 35$, in this case $dx/dt = -10$, which means that x will decrease as time progresses. As you may notice now, the differential equation, $dx/dt = 25 - x$, meaning that x will increase if x is less than 25, and will decrease if x is greater than 25. As a result, x will become 25 in future regardless of the current value of x. This kind of regulation might be employed in an air-conditioner to adjust the room temperature to a desired value (e.g., 25 °C).

6.1.2 Modeling a Biological Phenomenon Using Differential Equation

Once we model a phenomenon using a differential equation, we can predict the behavior of the phenomenon. The most important thing is how we can model a phenomenon, using a differential equation. As observed above, a simple differential equation $dx/dt = F(x)$ can describe the time (t)-dependent change in the value x. In this case, we need to formulate the change of x over time as a function of x.

For example, a logistic function,

$$dx/dt = rx(1 - x/K), \qquad (6.2)$$

is a classic model of population growth. Here, r and K are constants. Therefore, dx/dt will change depending only on x, which represents the population. This equation formulates two assumptions. First, the growth rate of the population (dx/dt) is basically proportional to the current population, x. This assumption is straightforward because the number of offspring should be large when the population is large. If this is the only assumption, the formulation will be $dx/dt = r \times x$, and the population will keep increasing infinitely. However, the second assumption preempts the infinite growth in population. It considers the limitation of resources, and considers that the population cannot grow if it approaches the value of K. This assumption is formulated in the part of $(1 - x/K)$. If x approaches K, $(1 - x/K)$ approaches 0. This means that $dx/dt = r \times (1 - x/K)$ also approaches 0.

An example from the field of molecular biology is the Michaelis-Menten equation. The equation is named after Leonor Michaelis and Maus Menten. The equation models the production rate of product (P), and the concentration of substrate (S), through an enzyme reaction. The equation is given as:

$$dP/dt = k_{cat} SX = k_{cat} X_T\{S/(K + S)\}, \qquad (6.3)$$

where k_{cat} is the catalytic rate of the enzyme reaction, SX and X_T are the concentrations of the enzyme-substrate complex and total enzyme, respectively, and K is the dissociate constant of the enzyme-substrate binding. The derivation of the Michaelis-Mentens equation is well explained in a textbook (Alon 2006).

6.2   What Differential Equations Convey

Differential equations convey the manner in which the variables in the equation change over time. To obtain complete information of such behaviors, we have to analytically solve the equation or numerically calculate the consequences, which are explained later in this chapter. These processes often require complicated calculation or development of a computer code. Before explaining how to obtain full information of a model expressed using differential equations, I will first explain how to obtain partial but important information about the model, without requiring complicated procedures.

6.2.1   Equilibrium Points

Differential equations describe the manner in which the variable, in the form of $dx/dt = F(x)$, changes over time. This means if $F(x) = 0$, then $dx/dt = 0$, and the amount of x will remain constant hereafter. The x that for which $F(x) = 0$ is known as the equilibrium point, and its value is obtained by solving the equation $F(x) = 0$.

Let us consider the logistic function introduced earlier in this chapter:

$$dx/dt = rx(1 - x/K). \tag{6.2}$$

Solving $dx/dt = 0$ gives a solution of $x = 0$, and K (when $r \neq 0$). Therefore, $x = 0$ and $x = K$ are the equilibrium points of the differential equation.

The consequence of the logistic function with various initial values are shown in Fig. 6.1 ($K = 10,000$, $r = 0.5$). The procedure for drawing such a graph is explained later, and drawing the logistic curve will be one of the objectives of this chapter. The lines shown in light blue (initial value 15,000), red (11,000), green (500), and blue (5), approach the value of K (10,000), and do not change further, after the value is attained. The violet line indicates the consequence of the logistic function, when the initial value is 0. Because violet line remains steady at 0, it means that $x = 0$ is an equilibrium point.

In summary, by solving the equation $dx/dt = 0$, which is far easier than solving the differential equation itself, we obtain information on the equilibrium points.

6.2.2   Stability of the Equilibrium Points: Linear Stability Analysis

In addition to the equilibrium points, we can easily determine the stability of the equilibrium points from the differential equation. Stability refers to whether the variable returns to the equilibrium points when a slight change of the value occurs at the equilibrium points.

In the above example (Eq. 6.2), the equilibrium points were $x = 0$, and K (10,000). To examine the stability of the equilibrium points, the sign of dx/dt is calculated, when $x = x* + \delta x$ ($x*$ is an equilibrium point, and δx is a small value). For $x* = 0$, $dx/dt > 0$

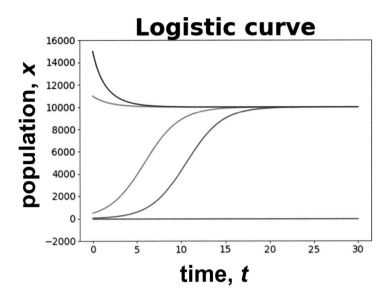

Fig. 6.1 The consequences of the logistic function

when $\delta x > 0$, and $dx/dt < 0$ when $\delta x < 0$. This means when x becomes slightly greater than $x^* = 0$, it will increase so that x is further away from 0. Similarly, when x becomes slightly less than 0, then x will decrease, and it will be further away from 0. As the slight change from $x^* = 0$ will not revert the value of x to 0, the equilibrium point of $x^* = 0$ is "unstable."

In contrast, for $x^* = K$, $dx/dt < 0$ when $\delta x > 0$, and $dx/dt > 0$ when $\delta x < 0$. This means, when x becomes greater than K, x will decrease and return to K. Similarly, when x becomes less than K, x will increase and return to K. Therefore, the equilibrium point of $x^* = K$ is "stable".

The linear stability analysis revealed that between two equilibrium points $x^* = 0$ and K, $x^* = 0$ is the unstable equilibrium point and $x^* = K$ is the stable equilibrium point. Overall, if $x > K$, x will decrease to approach K, and if $0 < x < K$, x will increase to approach K. Once x reaches K, then x is stable at K. If x is exactly 0, then x will not change. However, if x slightly changes from 0, x does not return to 0. Nevertheless, it continues to increase or decrease as $x = 0$ is an unstable equilibrium point. If $x < 0$, then x continues to decrease as $dx/dt < 0$ (Fig. 6.2). This overview is consistent with the behavior of the differential equation shown in Fig. 6.1, and thus we can obtain the qualitative behavior of the differential equation only by the linear stability analysis without solving the differential equation.

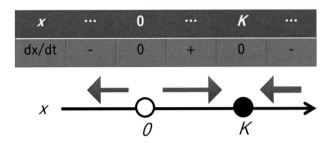

Fig. 6.2 Summary of the linear stability analyses of Eq. (6.2)

6.3   Solving Differential Equations

6.3.1   Modeling Nuclear Centration Using Differential Equation

As a simple example for modeling a biological process with differential equation, the nuclear centration is re-introduced. In Chap. 4, we modeled the process of nuclear centration, and simulated it with Microsoft Excel. We further conducted simulation with Python in Chap. 5. In these models, we calculated the velocity (v) of the nucleus from the current position (x) of the nucleus. For example, the relationship between the position and velocity of the nucleus in the cytoplasmic pulling model is shown below:

$$v = -\left(f_{\text{cyto}} cN/3\pi\eta R \right)x. \tag{4.3}$$

Most importantly, the velocity is the rate of change in the position, and thus, $v = dx/dt$. Therefore, (Eq. 4.3) can be rewritten as

$$dx/dt = -\left(f_{\text{cyto}} cN/3\pi\eta R \right)x. \tag{6.4}$$

This simple differential equation is sufficient to describe the cytoplasmic pulling model constructed in Chap. 4.

6.3.2   Analytical Solutions

Some differential equations can be solved analytically if they are simple. Solving a differential equation analytically means that the differential equation is converted into an equation without derivatives. Basically, if we consider the integral part of both sides of a differential equation, we can obtain an equation without derivatives. The techniques for solving differential equations analytically is out of the scope of this textbook. Please refer to specific math textbooks if you are interested in knowing them.

As a simple example, consider the differential equation of the cytoplasmic pulling model.

$$dx/dt = -\left(f_{\text{cyto}} cN/3\pi\eta R \right) x. \tag{6.4}$$

Because the parameters in $(f_{\text{cyto}} cN/3\pi\eta R)$ are all constants independent of x and t, a constant $k = (f_{\text{cyto}} cN/3\pi\eta R)$ is introduced. This makes Eq. (6.4) to $dx/dt = -k\,x$. A basic technique to solve differential equations is the separation of variables. In this case, we separate variables x and t to the left and right sides, respectively. Thus, the equation can be expressed as $(1/x)dx = -k\,dt$. Then the integral of both sides is taken as: $\int(1/x)dx = \int -k\,dt$. This gives $\log(x) = -kt + a$, where a is an integral constant. Furthermore, taking exponential of both sides gives us $x = A\,\exp(-kt)$, where $A = \exp(a)$. When $t = 0$, x will be equal to A. This means A is the initial position of x, and we rewrite $A = x_0$. In conclusion, the differential equation of Eq. (6.4) can be solved as follows:

$$x = x_0\,\exp\left\{-\left(f_{\text{cyto}} cN/3\pi\eta R \right) \times t\right\}. \tag{6.5}$$

Questions
1. Solve the logistic equation (Eq. 6.2) analytically. Let us assume $x = x_0$ at $t = 0$.

6.3.3 Calculating the Consequences of Differential Equations Computationally: Euler and the Runge–Kutta Methods

Once we have succeeded in solving a differential equation analytically, the behavior of the model will be known, by applying values to the solution. For example, the differential equation in Eq. (6.4) was solved as Eq. (6.5), and thus we know the behavior of Eq. (6.4) at a given time t.

However, in many cases, differential equations are difficult to solve analytically. Even in those cases, we are able to comprehend the numerical behavior of the equation by applying specific values to it. Because differential equations enable knowing the change from the current situation the near future will be known, and by repeating the procedure, the entire processes will be conveyed. The derivative is defined as:

$$\frac{dx}{dt} = \lim_{\Delta t \to 0} \frac{\Delta x}{\Delta t}. \tag{6.6}$$

When Δt is small, we can approximate,

$$dx/dt \fallingdotseq \Delta x/\Delta t = \{x(t + \Delta t) - x(t)\}/\Delta t. \tag{6.7}$$

A differential equation of

$$dx/dt = F(x) \tag{6.8}$$

can thus be approximated to

$$\{x(t + \Delta t) - x(t)\}/\Delta t = F(x), \tag{6.9}$$

which can be converted to

$$x(t + \Delta t) = x(t) + F(x)\Delta t. \tag{6.10}$$

This means the value of x in the near future $x(t + \Delta t)$ can be obtained from the value of $x(t)$ and $F(x)$. Note that this is an approximation, assuming Δt is sufficiently small. This method of obtaining the values of differential equation is known as the Euler method. This has been covered in Chap. 4 (with Microsoft Excel) as well as in Chap. 5 (with Python).

From the question at the end of Chap. 5, it can be inferred that this is only an approximation, and the obtained values depend on how small Δt is. Ideally Δt should be made indefinitely small to obtain accurate values. However, this is not realistic considering computational power. To increase the accuracy of obtaining numerical values from differential equations, another method known as the Runge–Kutta method, is more appropriate. Derivation of the method (Iserles 1996) is rather complicated and not explained here.

There are many variations of the Runge–Kutta methods. However, the classic Runge–Kutta method (RK4) is explained here. In the classic Runge–Kutta method, the consequence of the differential equation of (Eq. 6.8) is calculated as

$$x(t + \Delta t) = x(t) + \{(k_1 + 2k_2 + 2k_3 + k_4)/6\} \Delta t, \tag{6.11}$$

where

$$k_1 = F\{x(t)\},$$

$$k_2 = F\{x(t) + k_1(\Delta t/2)\},$$

$$k_3 = F\{x(t) + k_2(\Delta t/2)\},$$

$$k_4 = F\{x(t) + k_3\Delta t \}.$$

From the comparison of the Euler method (Eq. 6.10), and the classic Runge–Kutta method (Eq. 6.11), it can be inferred that in the classic Runge–Kutta method, the slope to calculate $x(t + \Delta t)$ from $x(t)$ is not simply $F(x)$ at $x = x(t)$. Instead, there is a weighted average of four kinds of $F(x)$ (i.e. k_1, k_2, k_3, and k_4). This makes the Runge–Kutta method provide a more accurate approximation of the differential equation compared to the Euler method.

Questions

2. Develop a python code to calculate the consequence of the logistic curve (Eq. 6.2) with the Euler method and check whether it agrees with Fig. 6.1.

6.3.4 A Coding Example of the Runge–Kutta Method with Python

Here, the consequence of the differential equation is calculated as follows (Eq. 6.4):

$$dx/dt = -\left(f_{cyto} cN/3\pi\eta R \right)x,$$

using the Runge–Kutta method (Fig. 6.3).

Code 6.1 Centration1D_CytoPull_RK4.py

```
"""
Centration1D_CytoPull_RK4.py
Created on Sun Jan 05, 2020
@author: akkimura (Python 3.5)
Description: simple 1D simulation for nuclear centration with Cytoplasmic
Pulling model only with the classic Runge-Kutta method (RK4)
"""
import numpy as np
from numpy import pi
import matplotlib.pyplot as plt


#%% Initialize parameters
L = 25 # half length of the cell [um]
N = 100 # number of microtubules (half)
R = 5 # the Stokes radius of the nucleus [um]
eta = 1 # the viscosity of the cytoplasm [pN s/um^2]
f = 1 # pulling force of a single motor [pN]
c = 0.01 # the density of the motors on microtubules [/um]
dt = 2 # sec per simulation step [s]
totSTEP = 120 # the number of simulation steps
X0 = 20 # initial position of the nucleus [um]
```

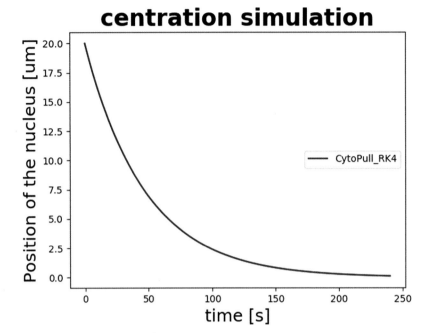

Fig. 6.3 Results of executing nuclear centration code "Centration1D_CytoPull_RK4.py"

```
#%% Initialize variables
Xcyto = np.zeros(totSTEP+1)
Xcyto[0] = X0


#%% Calculate
for st in range(totSTEP): # i starts from 0 to totSTEP-1
    # cytoplasmic pulling model, the classic Runge-Kutta method
    Coeff = -(f* c* N / (3* pi* eta* R))
    k1 = Coeff* Xcyto[st]
    k2 = Coeff* (Xcyto[st] + k1* dt / 2)
    k3 = Coeff* (Xcyto[st] + k2* dt / 2)
    k4 = Coeff* (Xcyto[st] + k3* dt)
    Xcyto[st+1] = Xcyto[st] + ((k1+2*k2+2*k3+k4)/6)* dt


#%% plot results
t_values = np.linspace(0,totSTEP*dt,totSTEP+1)
plt.plot(t_values, Xcyto, color="b", label="CytoPull_RK4")


# graph modifications
ax = plt.gca() # get current axis
ax.set_title("centration simulation", size=24, weight='bold')
ax.set_xlabel("time [s]", size=18)
```

```
ax.set_ylabel("Position of the nucleus [um]", size=18)
plt.legend(loc='center right') # show legends


#save the figure
fig = plt.gcf() # get current figure
plt.savefig("Centration1D_CytoPull_RK4.png")
```

Code 6.1 is mostly similar to Code 5.1 except that the former does not include calculations for the Cortical Pushing model or the Cortical Pulling model, and that Code 6.1 uses the classic Runge–Kutta method.

The difference in the codes for the classic Runge–Kutta method is demonstrated below:

```
# cytoplasmic pulling model, the classic Runge-Kutta method
Coeff = - (f* c* N / (3* pi* eta* R))
k1 = Coeff* Xcyto[st]
k2 = Coeff* (Xcyto[st] + k1* dt / 2)
k3 = Coeff* (Xcyto[st] + k2* dt / 2)
k4 = Coeff* (Xcyto[st] + k3* dt)
Xcyto[st+1] = Xcyto[st] + ((k1+k2/2+k3/2+k4)/6) * dt
```

For the Euler method, this part should be

```
# cytoplasmic pulling model, the Euler method
Coeff = - (f* c* N / (3* pi* eta* R))
k1 = Coeff* Xcyto[st]
Xcyto[st+1] = Xcyto[st] + k1* dt
```

When we compare the classic Runge–Kutta method, the Euler method, and the analytical solution (Fig. 6.4), it can be realized that the Runge–Kutta method is a better approximation to the differential equation compared to the Euler method.

Answers
1. $x(t) = K/\{1 + (K - x_0/x_0)\exp.(-r\,t)\}$. By separation of variables, we obtain $\{1/x + 1/(K - x)\}dx = r\,dt$. Integration of both sides yields $\log\{x/(K - x)\} = r\,t + c$, where c is an integral constant.
2. The following code is an example.

```
    """
    logistic.py
    Created on Wed May 11 15:38:57 2016
    @author: akkimura (Python 3.5)
    Description: calculate difference equation to draw a logistic curve
```

(continued)

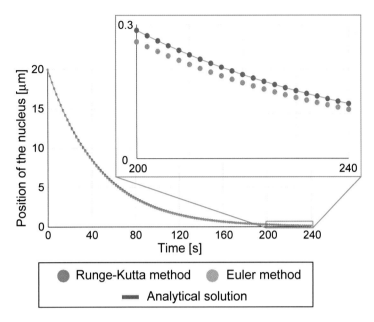

Fig. 6.4 Comparison between the classic Runge–Kutta method, the Euler method, and the analytical solution

```
"""
import numpy as np
import matplotlib.pyplot as plt

#%% Initialize parameters
P0 = 50
m = 0.5
K=10000
dt = 1
totSTEP = 30

#%% Initialize variables
deltaP = 0.0
P = np.zeros(totSTEP+1)
P[0] = P0

#%% Calculate
for i in range(totSTEP): # i starts from 0 to totSTEP-1
    deltaP = m * P[i] * (1-P[i]/K)
```

(continued)

```
    P[i+1] = P[i] + deltaP*dt

#%% plot results
t_values = np.linspace(0,totSTEP*dt,totSTEP+1)
plt.plot(t_values, P)

# graph modifications
ax = plt.gca()
ax.set_title("logistic curve", size=24, weight='bold')
ax.set_xlabel("time, t")
ax.set_ylabel("Population, x")

#save the figure
fig = plt.gcf()
fig.canvas.get_supported_filetypes()
plt.savefig("logistic_curve.png")
```

Take-Home Message
- Differential equation is a basic framework to model temporal changes of dynamic behavior.
- Linear stability analysis is an effective approach to understand the overall behavior of the model described with differential equations.
- Examples of numerical calculations coded with Python to know the behavior of differential equations are presented.

References

Alon U. An introduction to systems biology; 2006. https://doi.org/10.1201/9781420011432-9.
Iserles A. A first course in the numerical analysis of differential equations; 1996. https://doi.org/10.1017/cbo9780511995569.004.

Diversity of the Cell

7

Contents

What You Will Learn in This Chapter

Diversity is a key concept in biology that involves the discovery and description of diverse life forms. Finding commonalities among diverse life forms and structure is also important. In this chapter, I will provide examples using a common model to explain the diversity observed in cells. Examples include diversity in cell size, diversity due to fluctuations, diversity due to the environment, and diversity due to gene activity. Finally, I will discuss the role of genes in the context of the physical constraints of the cell.

7.1   Diversity of the Cell

Throughout the long history of natural selection, the existing species have always shown diversity. This also includes cells. There are various cell types, even in organisms that share common genetic information. Moreover, among the same cell type, individual cells are diverse in shape and sometimes function.

© Springer Nature Singapore Pte Ltd. 2022
A. Kimura, *Quantitative Biology*, Learning Materials in Biosciences,
https://doi.org/10.1007/978-981-16-5018-5_7

Science often focuses on commonalities and identifying unified mechanisms. However, the science of diversity is still in the development phase. I believe that quantitative modeling is a powerful approach for understanding diversity. As discussed in Chap. 1, a good model can "make reasonably accurate predictions". Thus, if a model accounts for the diversity in cell behavior, it can be regarded as a good model. Moreover, if a model can explain cellular diversity, we will be able to know the extent to which changes in assumptions or parameters account for the diversity in the model. This is an important step towards understanding the driving force behind diversity, which is possibly associated with changes in genetic information.

In this chapter, I will introduce some of my own research as an example of connecting the diversity of the cell with our mechanical models.

7.2   Diversity in Cell Size: Scaling Problems

Size is a fundamental feature of architecture. Therefore, regulations involving size are important in cell architectonics. The degree of diversity in cell size is small compared with that of organism size. The difference in organism size is thought to result from the number of cells rather than cell size. Nevertheless, there is also diversity in cell size. For example, in humans, eggs, which are spheres with diameters over 100 μm, are the largest cells, while other cells range in size from a few to tens of micrometers. Despite these differences in size, cells can perform similar functions, such as cell division.

Interestingly, the size of cellular components is often correlated with cell size. This correlative change in the size of cells and cellular components is called size scaling. In many cell types, it has been reported that the size of the nucleus is almost proportional to the size of the cell, i.e., the ratio between the nucleus and cytoplasm (N/C ratio) is conserved. The yeast *Schizosaccharomyces pombe* is a good example of this, where the N/C ratio is conserved even when the cell size is manipulated in mutants (Neumann and Nurse 2007). In one of our studies, we focused on the embryogenesis of the worm *Caenorhabditis elegans*, which has a transparent body and is thus useful for visualizing cellular structures inside the body. During embryogenesis, embryonic cells continue dividing while the total volume is conserved, as the cells are confined within the rigid eggshell. Therefore, the size of the cell gradually shrinks during embryogenesis. When we quantified the size of the intracellular structures, we found that nuclear, spindle, and chromosome size correlate with cell size (Hara and Kimura 2009, 2013; Hara et al. 2013). In most cases, the relationship was not isometric (perfectly proportional) but allometric (Hara and Kimura 2011).

In one of our research studies, we investigated size scaling in cells to connect cellular diversity with the mechanics of the cell (Hara and Kimura 2009). To this end, we first measured the relationship between cell size and spindle elongation size or speed during early embryogenesis of *C. elegans*. Given that a trimeric G protein and its regulators are involved in the elongation of the spindle, we also knocked down regulators of G proteins

(GPR-1 and 2) to inhibit spindle elongation. In the control (wild type), both the extent and speed of spindle elongation correlated with cell size. In the GPR-1/2-knockdown cells , the spindle elongated to a shorter extent than that of the wild type, but to an extent that was still correlated with cell size. In contrast, the speed of elongation in GPR-1/2-knockdown cells did not correlate with cell size. Therefore, we concluded that spindle elongation in wild-type cells consists of at least two mechanisms that are dependent and independent of GPR-1/2 proteins. Subsequently, we constructed two hypothetical models to account for both dependent and independent mechanisms. Numerical simulation based on the models was able to explain the quantitative features of the experimental results. Therefore, the models constructed in the study can account for cellular diversity with regards to cell size and spindle elongation.

7.3   Diversity in Cellular Response Due to Fluctuations

The inside of the cell is full of noise. Molecules and their complexes randomly move with thermal fluctuations and in an energy (ATP)-dependent manner everywhere inside the cell (see Chap. 8). Gene expression involves numerous stochastic processes. As a result, the number and concentration of a particular molecule fluctuates. Such fluctuations sometimes alter cell fate. Small fluctuations can be amplified through positive feedback regulation (see Chap. 10), resulting in critical consequences both in vivo (Greenwald and Rubin 1992) and in artificial systems (Elowitz et al. 2002).

Even without considering fancy mechanisms to amplify the fluctuations, diversity is still observed. Let us consider the simple chemical (binding and dissociation) reactions of two molecules, A and B:

$$A + B \rightleftharpoons AB$$

At equilibrium, A is in a "diverse" form, either alone (A) or in complex with B (AB). The coexistence of multiple states is often explained by an energy landscape (Fig. 7.1). In general, states with lower potential energies are considered stable and favorable. This means that with an energy supply, a system can adopt states other than the minimum potential energy, exhibiting diverse states.

I will now introduce an elegant study that applied the energy landscape theory to explain the diverse behavior of the cell (Théry et al. 2007). The mitotic spindle is a molecular machine required to segregate sister chromatids into daughter cells before cytokinesis. The orientation of the spindle is important because it defines the plane of cell division. This orientation is thought to be determined mainly by the forces acting on the two poles of the spindle. Based on the accumulated knowledge on spindle orientation, the authors of the paper constructed a mechanical model to calculate the forces acting on the spindle pole. Once the force was calculated, the authors were able to calculate the potential energy of a given spindle orientation. The difference in potential energy here corresponds to the work

Fig. 7.1 A diagram of the
energy landscape

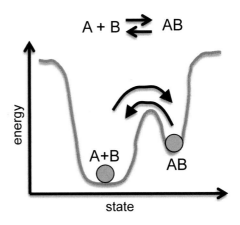

required to rotate the spindle from one orientation to another. Thus, the authors were able to determine the energy landscape for spindle orientation. A further interesting aspect of this study is that the authors were able to characterize the appearance frequency of various orientations. As the authors were able to precisely control the shape of the cell in their experiments, and thus force distribution, they were able to conclude that the spindle shows diverse orientations even under the same conditions. The frequency distribution of the spindle orientation characterized experimentally agreed well with the predicted frequency from the energy landscape (see Chap. 8 for the conversion of energy to frequency). Therefore, the mechanical model accounts for the diverse status of the cell and the frequency of orientations of the spindle.

We applied this approach to understand the number of division of a mutant cell (Kondo and Kimura 2019). The mitotic spindle is usually a bipolar structure, with the two centrosomes representing the two poles. In the mutant cell of a *C. elegans* embryo, we observed three or more centrosomes due to a dysregulation in centrosome number. Accordingly, with three centrosomes, a tripolar—instead of bipolar—spindle forms. Because the spindle defines the cell division plane, the tripolar spindle induces cell division into three daughter cells. In other words, the tripolar spindle induces two cleavage furrows. Interestingly, the tripolar spindle did not always induce two cleavage furrows, as cytokinesis with one cleavage furrow was also observed. In fact, only about a third of the cells with tripolar spindles showed two cleavage furrows, whereas the rest had one furrow. To understand the mechanism underlying this behavior, we analyzed the orientation of the tripolar spindle using a mechanical model similar to that introduced in the previous paragraph (Théry et al. 2007). We were able to show that the number of cleavage furrows correlated with the orientation of the tripolar spindle and that the frequency distribution of the orientation was well explained by the energy landscape calculated from the model. In conclusion, the energy landscape can be used to account for the diversity of cell behavior based on a mechanical model of the cell.

7.4   Diversity in Cell Arrangement Due to Spatial Restrictions

Within a cell population, cells must position themselves. This positional arrangement is important as it defines the neighboring cells for adherence, which affects the type of chemical signals exchanged. Such adjacency relation is particularly important for embryonic cells as the cells determine their fate depending on signaling with neighboring cells.

Patterns of cell arrangement are diverse, even among related species. Among nematode species, for example, four distinct patterns of cell arrangement are observed at the 4-cell stage that are referred to as pyramid, diamond, T-shape, and linear type (Fig. 7.2). Several factors, such as the orientation or timing of cell division, may influence cell arrangement. In our research, we focused on the shape of the eggshells (Yamamoto and Kimura 2017); when we compared the arrangement and the eggshell shape, we found that the pyramid-type pattern is observed in the spherical eggshell, while the diamond type is observed as the eggshell elongates to become ellipsoidal. The T-shape and linear types are observed when the eggshell is further elongated. Using the *C. elegans* embryo, we were able to demonstrate that changing the shape of the eggshell alters the cell arrangement. All four arrangement types and their dependency on sphericity of the eggshell were successfully reproduced by the model we created by modifying a previously proposed mechanical model (Fickentscher et al. 2013). Therefore, our study (Yamamoto and Kimura 2017) supports the notion that the diversity in cell arrangement can be explained by spatial confinement within the eggshell.

7.5   Diversity in the Pattern of Cytoplasmic Streaming Due to Molecular Activities

The final example in this chapter involves mechanical modeling of the diversity in cytoplasmic streaming. Cytoplasmic streaming—the massive flow of the cytoplasm inside the cell—is thought to contribute to the transport of materials within the cell. In some cases, the direction of streaming is defined, possibly due to the polarity of the cell. In other cases, the direction is not predefined. About half a century ago, Noburo Kamiya classified the patterns of cytoplasmic streaming into seven classes (Kamiya 1959); among them, four classes show predetermined directions, whereas for the other three, the direction is not predetermined. The latter three classes include (1) saltation, (2) circulation, and (3) rotation. Saltation involves local cytoplasm flow, without an overall coordinated direction. For the circulation type, the overall direction is coordinated, but the direction changes from time to time. Finally, in rotation, the overall direction is almost fixed (but not predetermined) throughout cytoplasmic streaming.

We became interested in this topic as we observed circulation-type streaming in the zygotes of our favorite model organism, *C. elegans*. We refer to this streaming as meiotic cytoplasmic streaming, which has been described and characterized by Francis McNally's group (McNally et al. 2010). We were particularly interested in the fact that the direction of

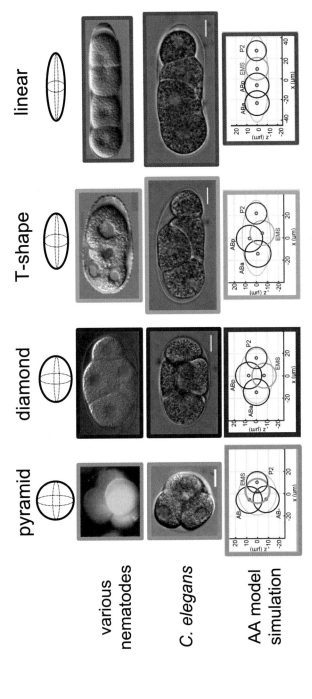

Fig. 7.2 Distinct arrangement of nematode embryo at the 4-cell stage. (Top) various nematode species show distinct cell arrangements at the 4-cell stage. The patterns of *Enoplus brevis* (pink, image from Schulze and Schierenberg (2011)), *C. elegans* (blue, image from Yamamoto and Kimura (2017)), *Cephalobus* sp. (green, image from Goldstein (2001)), and *Belonolaimus* sp. (purple, image from Goldstein (2001)) are shown. (Middle) By changing the aspect ratio of the ellipsoidal eggshell of the *C. elegans* embryo, we were able to reproduce the patterns of other species (Yamamoto and Kimura 2017). (Bottom) A computational model that reproduces the diversity and robustness of the cell arrangement patterns (Yamamoto and Kimura 2017)

the streaming changes occasionally, and constructed a numerical model to describe its emergence and reversal (Kimura et al. 2017) (see also Chap. 9). Interestingly, the model not only reproduced circulation-type streaming, but also saltation- and rotation-type streaming by changing the values of the model parameters (Fig. 7.3). Moreover, we were able to convert the circulation-type streaming observed in the cell into saltation- and rotation-type streaming by manipulating gene expression, as predicted from the model. When we increased the stability of microtubules by inhibiting the enzyme responsible for severing microtubules, rotation-type streaming was observed. Meanwhile, when the connectivity of the network structures of the endoplasmic reticulum (ER) was impaired, saltation-type streaming was observed. These findings demonstrate that diverse cellular behaviors can be explained by a common model where the parameters are changed to correspond to gene functions. Thus, the quantitative model plays a critical role in understanding the diverse behaviors of the cell.

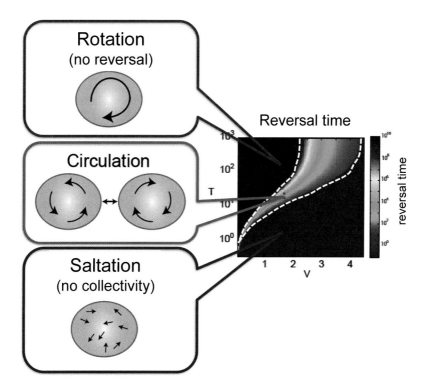

Fig. 7.3 Distinct types of cytoplasmic streaming reproduced using a common model. By changing the parameters in a model originally constructed for circulation-type cytoplasmic streaming of the *C. elegans* zygote (Kimura et al. 2017), we can reproduce the rotation- and saltation-type streaming observed in other species (Kamiya 1959)

7.6 The Role of a Gene as a Switch

As I studied cell architectonics, one big question was to what extent do genes define cell behavior and their exact function. It was hard to believe that genes define every detail of what a cell should do. Mechanical modeling tells us that simple models based on mechanical activities and constraints can explain not only a particular status/behavior of the cell, but also diverse statuses and behaviors. Drs. Šiber and Ziherl argued in their book *Cellular Patterns* that *"[the] genes and the proteins that they encode appear only as regulators of the physical necessity, switching the flow of matter and energy at the critical points in space and time from one physical possibility to the other. (. . .) In some sense, biology and evolution "explore" physics"* (p. 4–5 in (Šiber and Ziherl 2017)). I favor this idea of viewing gene function as a switch. Thus, understanding the interplay between physics and genetics is important.

Take-Home Message
- The diverse behaviors of the cell can be explained by a common model with different parameter values or fluctuations.
- Genes may act as switches to select among limited choices created by physical constraints.

References

Elowitz MB, Levine AJ, Siggia ED, Swain PS. Stochastic gene expression in a single cell. Science. 2002;297:1183–6.

Fickentscher R, Struntz P, Weiss M. Mechanical cues in the early embryogenesis of *Caenorhabditis elegans*. Biophys J. 2013;105:1805–11.

Goldstein B. On the evolution of early development in the Nematoda. Philos Trans R Soc Lond Ser B Biol Sci. 2001;356:1521–31.

Greenwald I, Rubin GM. Making a difference: the role of cell-cell interactions in establishing separate identities for equivalent cells. Cell. 1992;68:271–81.

Hara Y, Kimura A. Cell-size-dependent spindle elongation in the *Caenorhabditis elegans* early embryo. Curr Biol. 2009;19:1549–54.

Hara Y, Kimura A. Cell-size-dependent control of organelle sizes during development. Results Probl Cell Differ. 2011;53:93–108.

Hara Y, Kimura A. An allometric relationship between mitotic spindle width, spindle length, and ploidy in Caenorhabditis elegans embryos. Mol Biol Cell. 2013;24:1411–9.

Hara Y, Iwabuchi M, Ohsumi K, Kimura A. Intranuclear DNA density affects chromosome condensation in metazoans. Mol Biol Cell. 2013;24:2442–53.

Kamiya N. Protoplasmic streaming. Springer; 1959.

Kimura K, Mamane A, Sasaki T, et al. Endoplasmic-reticulum-mediated microtubule alignment governs cytoplasmic streaming. Nat Cell Biol. 2017;19:399–406.

Kondo T, Kimura A. Choice between 1- and 2-furrow cytokinesis in *Caenorhabditis elegans* embryos with tripolar spindles. Mol Biol Cell. 2019;30:2065–75.

McNally KL, Martin JL, Ellefson M, McNally FJ. Kinesin-dependent transport results in polarized migration of the nucleus in oocytes and inward movement of yolk granules in meiotic embryos. Dev Biol. 2010;339:126–40.

Neumann FR, Nurse P. Nuclear size control in fission yeast. J Cell Biol. 2007;179:593–600.

Schulze J, Schierenberg E. Evolution of embryonic development in nematodes. EvoDevo. 2011;2:18.

Šiber A, Ziherl P. Cellular patterns. CRC Press; 2017.

Théry M, Jiménez-Dalmaroni A, Racine V, Bornens M, Jülicher F. Experimental and theoretical study of mitotic spindle orientation. Nature. 2007;447:493–6.

Yamamoto K, Kimura A. An asymmetric attraction model for the diversity and robustness of cell arrangement in nematodes. Development. 2017;144:4437–49.

Randomness, Diffusion, and Probability

8

Contents

What You Will Learn in This Chapter

Biological systems are not static, uniform, or deterministic, but rather dynamic, heterogeneous, and probabilistic. To understand and model such systems, randomness and related concepts are important. In this chapter, I will first introduce randomness and how to construct models involving randomness. Next, I will introduce diffusion as a consequence of random movements. Finally, the Boltzmann distribution is introduced as a consequence of randomness. Boltzmann distribution is important when we want to calculate the probability of stochastic phenomena.

Learning Objectives

After completing this chapter, readers should be able to

© Springer Nature Singapore Pte Ltd. 2022   85
A. Kimura, *Quantitative Biology*, Learning Materials in Biosciences,
https://doi.org/10.1007/978-981-16-5018-5_8

1. Create computational codes with random numbers (stochastic simulations).
2. Understand the concept and formulation of diffusion.
3. Understand the concepts of "energy landscape" and Boltzman distribution.

Important Concepts Discussed in This Chapter

- *Diffusion*; small molecules move easily, even with thermal fluctuations. Random movements make a population of molecule diffuse from the source. Diffusion is important to describe the behavior of molecules floating inside the cell.
- *Boltzmann distribution*; distribution of probability in the context of a certain energy.

8.1 Randomness

8.1.1 Why Should We Consider Randomness for Biological Processes?

Biological molecules are always moving inside and outside the cell. A driving force for such movement is thermal energy: every molecule at an absolute temperature T has the thermal energy of $k_B \times T/2$ per degree-of-freedom, where k_B is the Boltzmann's constant (1.38×10^{-23} J/K). At our body temperature of 310 [K] (~37 °C/~98 °F), the thermal energy will be ~2×10^{-21} J. How large is this? According to Newton's law, the kinetic energy of a particle with its mass of m and velocity of v is $mv^2/2$. When we consider a lysozyme protein (molecular weight = 14,000) as an example of biological molecules, the velocity will be $v = (k_B T/m)^{0.5} =$ ~10 [m/s], which is very fast. Inside the cell, every molecule collides with other molecules and changes direction; thus, a molecule does not move unidirectionally for meters in a second. Still, every molecule is vigorously moving in random patterns inside the cell. In addition to thermal energy, molecules inside the cell use metabolic energy to move; one molecule will collide with surrounding molecules, thus influencing molecular movement. Certain molecules may move with defined direction, but most move randomly, indicating the importance of random movement.

8.1.2 Modeling Random Motion with Python

Randomness is like flipping a coin. The chance of getting heads or tails is random with the same probability. To create random numbers in Python, we use a module named "numpy. random" from the "numpy" library. We can create a random number between 0 and 1 with the `numpy.random` function. `numpy.random()` is a random number, and `numpy. random(10)` is an array with 10 random numbers. For example, the following is a code to create 2-dimensional random motion (Fig. 8.1).

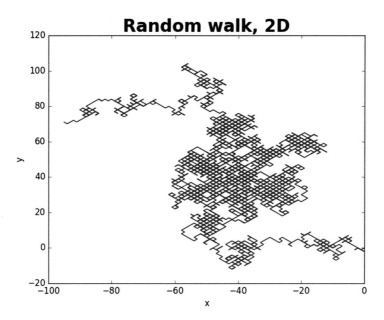

Fig. 8.1 The result of running Code 8.1

Code 8.1 RandomWalk2D.py

```
"""
Created on Sat May 21 11:15:17 2016
@author: akkimura (Python 3.5)
Description:generate random numbers and conduct random walk in 2D
"""

import numpy as np
import matplotlib.pyplot as plt
from numpy.random import random as rng
num_steps = 5000

t_array = np.arange(num_steps)
x = np.empty(num_steps)
y = np.empty(num_steps)
r_array = rng((num_steps,2))

for i in t_array:
  if i==0:
    x[i]=0
    y[i]=0
```

```
else:
  if r_array[i,0]>0.5:
    x[i]=x[i-1]+1
  else:
    x[i]=x[i-1]-1
  if r_array[i,1]>0.5:
    y[i]=y[i-1]+1
  else:
    y[i]=y[i-1]-1

plt.plot(x,y)

# graph modifications
ax = plt.gca()
ax.set_title("Random walk, 2D", size=24, weight='bold')
ax.set_xlabel("x")
ax.set_ylabel("y")

#save the figure
fig = plt.gcf()
fig.canvas.get_supported_filetypes()
plt.savefig("RandomWalk2D.png")
```

For each step, we create two random numbers between 0 and 1. In our example, numpy.random is abbreviated as rng (as shown in the second row of the following extracted code). In the fourth row, rng((num_steps,2)) means a 5000 × 2-sized matrix of random numbers (between 0 and 1).

```
import numpy as np
from numpy.random import random as rng
num_steps = 5000
r_array = rng((num_steps,2))
```

If the first random number is larger than 0.5, the particle moves one step towards the plus direction along the x-axis, and if less than 0.5, the particle moves towards the minus. In the code, we perform the same function for the y-axis, using the second random number. We realize these conditional branching with if (condition): else:

```
if r_array[i,0]>0.5:
  x[i]=x[i-1]+1
else:
  x[i]=x[i-1]-1
```

8.2   Diffusion

8.2.1   Random Motion and Diffusion

Next, we consider 1-dimensional random movement of many particles starting from the position $x = 0$ (at time $= 0$). At the next step, let us move 30% to $x = +1$, another 30% to $x = -1$, while the rest (40%) remain at 0. Let us repeat this procedure. For example, after the second step, 9% will be at +2, and another 9% will be at -2. This procedure can be generalized as follows: The concentration of the particle at $x = i$ position at $t = j$ time, C [i, j], follows the relationship indicated herein:

$$C[i, j + 1] = C[i,j] + 0.3 \times \{C[i + 1,j] + C[i - 1,j]\} - 0.6 \times C[i,j]. \tag{8.1}$$

Code 8.2 is a Python code to repeat the procedure multiple times (as in the example, $\texttt{step} = 100$).

Code 8.2 Diffusion.py

```
# -*- coding: utf-8 -*-
"""
Created on Sat May 21 16:10:34 2016
@author: akkimura (Python 3.5)
Description:
"""

import numpy as np
import matplotlib.pyplot as plt

step = 100
x_array = np.arange(-step, step+1, 1)
C = np.zeros((2*step+1,step+1))
Cinit = 1.0
C[step,0] = Cinit

# diffusion coefficient (non-dimension)
df = 0.3

for t in np.arange(step):
  for x in np.arange(2*step+1):
    if x==0:
      C[x,t+1] = C[x,t] + df*C[x+1,t] - df*C[x,t]
    elif x==2*step:
      C[x,t+1] = C[x,t] + df*C[x-1,t] - df*C[x,t]
    else:
      C[x,t+1] = C[x,t] + df*C[x+1,t] + df*C[x-1,t] - 2*df*C[x,t]
```

```
# draw graph
plt.figure(1)
samplingST = [1,10,20,40,100]
for i in samplingST:
  plt.plot(x_array,C[:,i])
ax = plt.gca()
ax.set_title("Diffusion", size=24, weight='bold')
ax.set_xlabel("x", size=18, weight='bold')
ax.set_ylabel("freq", size=18, weight='bold')
fig = plt.gcf()
fig.canvas.get_supported_filetypes()
plt.savefig("Diffusion.png")


# MSD calculation
MSD = np.zeros(step+1)
for t in np.arange(step+1):
  SumSD = 0.0
  Count = 0.0
  for x in np.arange(2*step+1):
    SumSD = SumSD + np.power(x_array[x],2)*C[x,t]
    Count = Count + C[x,t]
  MSD[t] = SumSD/Count


plt.figure(2)
t_array = np.arange(step+1)
plt.plot(t_array,MSD)
ax2 = plt.gca()
ax2.set_title("Diffusion", size=24, weight='bold')
ax2.set_xlabel("t", size=18, weight='bold')
ax2.set_ylabel("MSD", size=18, weight='bold')
fig2 = plt.gcf()
fig2.canvas.get_supported_filetypes()
plt.savefig("Diffusion_MSD.png")
```

In the code, we made the degree of movement at one-step adjustable by introducing a parameter, df. If df = 0.3, the 30% goes to +1 and another 30% goes to -1 positions (i.e., loose 60% in total). In the code, we prepare a matrix C with the size of $(2 \times$ step $+ 1) \times ($step $+ 1)$. As the range of x position if from 0 to $2 \times$ step, there is no $C[i - 1, j]$ for $i = 1$ nor $C[i + 1, j]$ for $i = 2 \times$ step. For this section, we used if: elif: else: statement.

```
for x in np.arange(2*step+1):
  if x==0:
    C[x,t+1] = C[x,t] + df*C[x+1,t] - df*C[x,t]
  elif x==2*step:
    C[x,t+1] = C[x,t] + df*C[x-1,t] - df*C[x,t]
  else:
    C[x,t+1] = C[x,t] + df*C[x+1,t] + df*C[x-1,t] - 2*df*C[x,t]
```

The distribution of the particle position at different steps was visualized by plotting $C[i, j]$ against x at time $t = j$, as in Fig. 8.2. The range covered by the particles will be larger as time progresses. Importantly, the average position is always at $x = 0$ because the distribution is symmetrical against $x = 0$. In addition, the mode (peak of the histogram) is also $x = 0$. This random spreading from the initial point (i.e., source) is called diffusion.

Since the average position does not change as time progresses, we require an index other than the average position to evaluate the degree of the movement. Mean square displacement (commonly abbreviated as MSD) is used for this purpose. As the displacement is squared in calculation of MSD, the value will increase as time progresses. In Code 8.2, we also calculate MSD. numpy.power(x,n) calculates n-th power of x. numpy.average is a function to calculate the average. Figure 8.3 plots MSD against time. For pure random movement (i.e., Brownian motion), MSD is proportional to the elapsed time.

Diffusion is an important movement type in biological systems. Inside the cell, not every movement molecular movement is driven by motor proteins gliding along

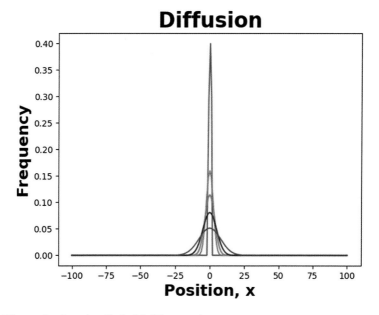

Fig. 8.2 The result of running Code 8.2 (histogram)

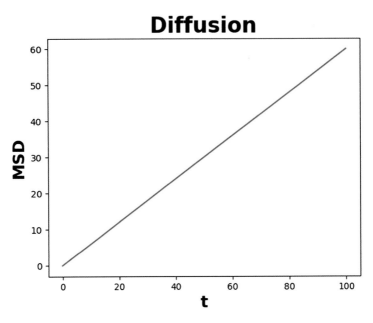

Fig. 8.3 The result of running Code 8.2 (MSD)

cytoskeletons. Most molecules use diffusion for delivery into the cell, in order to encounter their target molecules (e.g., signal transduction).

Another example is the formation of morphogen gradient. The specification of cell fate is important during development. Morphogens are molecules defining cell fate depending on its concentration. Morphogen is said to induce a gradient across tissues. The concentration near the source of the morphogen is high, but an area further away from the source exhibits a lower concentration. This is due to diffusion. The difference in the concentration function as a positional cue to measure the distance from the source and to generate cells with different fates is distant-dependent.

8.2.2   Diffusion Equation

In Chap. 6, we modeled the temporal change of a biological system with differential equations. To incorporate diffusion to the models described with differential equations, we will try to formulate diffusion mathematically.

Here, I consider a one-dimensional discrete model identical to how we constructed Code 8.2 above. $C(x,t)$ is the concentration of the material at position x and time t. $2 \times D/\Delta x^2$ is the proportion moving from position x during one time step (Δt). The concentration at x at the next step ($t + \Delta t$) will be:

$$C(x, t + \Delta t) = C(x, t) + (D/\Delta x^2) \times [-2C(x, t) + C(x - \Delta x, t) + C(x + \Delta x, t)]$$
$$\times \Delta t. \tag{8.2}$$

During one step, $2 \times (D/\Delta x^2) \times C(x,t) \times \Delta t$ will move away from x, but $(D/\Delta x^2) \times C(x - \Delta x,t) \times \Delta t$ and $(D/\Delta x^2) \times C(x + \Delta x,t) \times \Delta t$ will move into x from $x - \Delta x$ and $x + \Delta x$ positions, respectively. Equation (8.2) will be converted to

$$\frac{C(x, t + \Delta t) - C(x, t)}{\Delta t} = D \times \left[\frac{C(x + \Delta x, t) - C(x, t)}{\Delta x} - \frac{C(x, t) - C(x - \Delta x, t)}{\Delta x} \right] / \Delta x \tag{8.3}$$

The definition of the derivative (Chap. 6), was $\frac{dF(t)}{dt} = \lim_{\Delta t \to 0} \frac{F(t+\Delta t)-F(t)}{\Delta t}$, when F is the function of one variable, t. For a function of multiple variables (e.g., $F(x,t)$), the partial derivative is defined as $\frac{\partial F(x, t)}{\partial t} = \lim_{\Delta t \to 0} \frac{F(x, t+\Delta t)-F(x, t)}{\Delta t}$. Using partial derivatives and the limit, Eq. (8.3) will be

$$\frac{\partial C(x, t)}{\partial t} = D \frac{\partial^2}{\partial x^2} C(x, t) \tag{8.4}$$

The three-dimensional version of Eq. (8.4) will be

$$\frac{\partial C(x, y, z, t)}{\partial t} = D \left\{ \frac{\partial^2 C}{\partial x^2} + \frac{\partial^2 C}{\partial y^2} + \frac{\partial^2 C}{\partial z^2} \right\} = D\nabla^2 C(x, y, z, t) \tag{8.5}$$

where ∇ (nabla) is the vector differential operator defined as $\nabla = \left(\frac{\partial}{\partial x}, \frac{\partial}{\partial y}, \frac{\partial}{\partial z} \right) = \vec{x} \frac{\partial}{\partial x} + \vec{y} \frac{\partial}{\partial y} + \vec{z} \frac{\partial}{\partial z}$.

8.3   Energy Landscape and Existing Probability

8.3.1   Potential Energy and Energy Landscape

As a related topic to randomness, I would like to mention energy landscape and probability. If you release a ball on a slope (Fig. 8.4(i)), the ball will roll down the slope and stop at the bottom, because of gravity. You might learn in a physics course that an object at a higher position has higher (gravitational) potential energy, and an object is stable at a position with lower potential energy. This is an example of the minimum total potential energy principle, in which the state that (locally) minimized the total potential energy will be the most stable state. When we consider another landscape, such as Fig. 8.4(ii), and release the ball at position "b", the ball will reach the bottom of well b. While the bottom of the well a, is

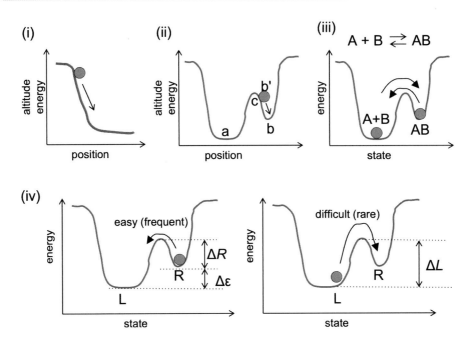

Fig. 8.4 Potential energy landscape

lower than that of *b*, the ball will remain in *b* because it is difficult to climb up the wall *c*, between *a* and *b*. In other words, if the ball has enough (kinetic) energy to climb up the wall *c*, the ball at *b* will move to *a*. Therefore, the probability of the ball to move from *b* to *a* (or *a* to *b*) is a matter of the kinetic possessed by the ball.

Another example is a chemical reaction. Consider a simple reversible reaction that component A and B associate and dissociate, we formulate the reaction as in Fig. 8.4(iii). The reversible reaction means that, even the bound form of A+B is more stable (lower energy) than the dissociated form (or vice versa), and both forms co-exist. This means that the "wall" (or energy barrier) between the two states is relatively low compared to the kinetic energy the molecules have.

What is the kinetic energy that molecules possess? In the introduction of this chapter, I mentioned thermal energy. Every molecule at absolute temperature T has a kinetic energy of $k_B T/2$ (per degree-of-freedom), on average. k_B is the Boltzman's constant (1.38×10^{-23} J/K). From this statement, we know the average thermal energy a molecule has, but, considering the movement across the energy barrier, what proportion of the molecule has large enough energy? In other words, what is the distribution of kinetic energy?

8.3.2   Boltzmann Distribution

The Boltzmann distribution is a probability distribution of a state with a certain energy and is expressed as follows:

$$p_i \propto \exp\left(-\frac{E_i}{k_B T}\right),$$

where p_i is the probability of containing energy E_i. This means that the probability will decrease exponentially as the energy increases. Figure 8.5 shows curves with exponential decay.

Why does the distribution follow this form? Fumio Oosawa, a biophysicist known as father of biophysics in Japan, explained this distribution to a lay person with simple experiments using a dice (Oosawa 2011). Six people, initially with the same number of coins, play a simple game with the following rule. At each step, one throws the dice twice. Suppose the procedure gave the numbers 4 and 2, in this order. Then, player 4 gives one coin from his stock to player 2. This step is repeated several times, and the distribution of the coins each player has will follow the Boltzmann distribution. This game can mimic the situation of molecules repeatedly colliding, and transferring kinetic energy from one to the other, in the gas state.

Code 8.3 is a Python version of Oosawa's game. In this case, we set the number of players, `ParticleN`, to 100 and repeat for `TotStep2` = 1,000,000 steps.

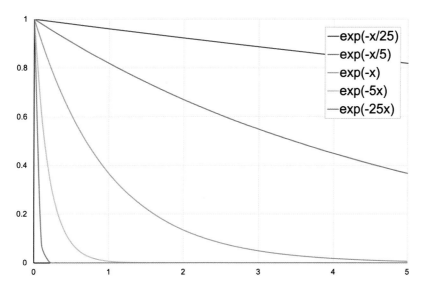

Fig. 8.5 Curves depicting exponential decay

Code 8.3 Boltzmann.py

```python
# -*- coding: utf-8 -*-
"""
Created on Fri May 27 16:50:30 2016
Revised on Sat Apr 17 2021
@author: akkimura (Python 3.7)
Description: an Osawa model to understand Boltzmann distribution
"""

import numpy as np
import matplotlib.pyplot as plt
from numpy.random import random as rng

ParticleN = 100
Score = np.zeros(ParticleN)
TotStep1 = ParticleN*100
TotStep2 = ParticleN*10000
RandN1 = rng(TotStep1)
RandN2 = rng((TotStep2,2))
Dice1 = np.floor(RandN1*ParticleN)
Dice2 = np.floor(RandN2*ParticleN)

#%% share energy
for st1 in np.arange(TotStep1):
    temp1 = int(Dice1[st1])
    Score[temp1] = Score[temp1]+1

#%% draw graph
plt.figure(1)
plt.hist(Score)
ax1 = plt.gca()
ax1.set_title("Boltzmann_share", size=24, weight='bold')
ax1.set_xlabel("Score", size=18, weight='bold')
ax1.set_ylabel("Count", size=18, weight='bold')
fig1 = plt.gcf()
fig1.canvas.get_supported_filetypes()
plt.savefig("Boltzmann_share.png")
plt.savefig("Boltzmann_share.tif")

#%% exchange energy
for st2 in np.arange(TotStep2):
    temp20 = int(Dice2[st2,0])
    temp21 = int(Dice2[st2,1])
    if Score[temp20] > 0:
```

```
    Score[temp20] = Score[temp20]-1
    Score[temp21] = Score[temp21]+1

#%% draw graph
plt.figure(2)
plt.hist(Score)
ax2 = plt.gca()
ax2.set_title("Boltzmann_exchange", size=24, weight='bold')
ax2.set_xlabel("Score", size=18, weight='bold')
ax2.set_ylabel("Count", size=18, weight='bold')
fig2 = plt.gcf()
fig2.canvas.get_supported_filetypes()
plt.savefig("Boltzmann_exchange.png")
plt.savefig("Boltzmann_exchange.tif")
```

Figure 8.6 exhibits the result obtained by running the code. We obtain a similar distribution to the Boltzmann distribution.

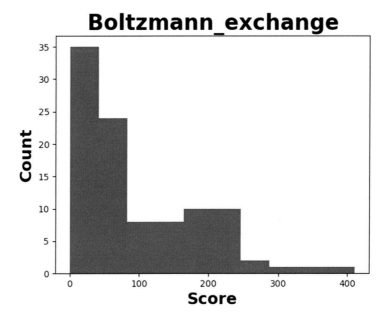

Fig. 8.6 A result of running Code 8.3

8.3.3   Existing Probability

The interesting point about the Boltzmann distribution is that it can connect differences in (potential) energy and the probability of the state. In the example depicted in Fig. 8.4(iv), to change from the right state to the left requires energy greater than ΔR, which will be $\int_{\Delta R}^{\infty} E_0 \exp\left(-\frac{x}{k_B T}\right) dx = -E_0 k_B T \left[\exp\left(-\frac{x}{k_B T}\right)\right]_{x=\Delta R}^{x=\infty} = E_0 k_B T \exp\left(-\frac{\Delta R}{k_B T}\right)$. Here, E_0 is the proportional constant. Similarly, the proportion of molecules moving from the left to right is $E_0 k_B T \exp\left(-\frac{\Delta L}{k_B T}\right)$. At equilibrium, molecules moving from right to left equals those moving from left to right. Thus,

$$[R] \times E_0 k_B T \exp\left(-\frac{\Delta R}{k_B T}\right) = [L] \times E_0 k_B T \exp\left(-\frac{\Delta L}{k_B T}\right). \tag{8.6}$$

From this equation,

$$\frac{[R]}{[L]} = \frac{\exp\left(-\frac{\Delta L}{k_B T}\right)}{\exp\left(-\frac{\Delta R}{k_B T}\right)} = \exp\left\{-\frac{(\Delta L - \Delta R)}{k_B T}\right\}. \tag{8.7}$$

This means that the ratio of existence depends on how large the difference in potential energy measures, normalized to the unit of energy (i.e., $k_B T$ in the case of thermal energy).

This concept can be, and has been, applied to scenarios in cell biology. Cellular behavior is not always deterministic, but sometimes stochastic. Such stochastic behavior can be explained by calculating potential energy and applying the Boltzmann distribution concept. As and example, I revisit the researches introduced in Chap. 7. The orientation of mitotic spindles, a spindle-shaped macromolecular assembly to segregate sister chromatids to the daughter cells, is critical to determine the orientation of the cell division plane. For a cultured cell line, the orientation of the spindle is thought to be determined by the torque generated by mechanical forces pulling the two poles from the outside of the spindle. Théry and colleagues constructed a physical model to calculate the torque applied to the spindle, and from the torque to calculate the potential energy each orientation of the spindle has in different geometrical environments. The calculated potential energy was then converted into the probability of the orientation. On the other hand, the authors experimentally determined the probability of the orientation in different environments. The probability calculated from the theory agreed qualitatively well with the experimental one. Moreover, by adjusting the amount of energy, they were able to show that there are appropriate units of energy that quantitatively agree with each environment (Théry et al. 2007). My colleagues and I applied this approach to explain the spindle orientation in other cases (Matsumura et al. 2016; Kondo and Kimura 2019).

Questions
1. Change the `df` constant in Code 8.2 and observe the results.
2. Construct a code to calculate the MSD of a two-dimensional random walk.
3. Change the parameters or rules in Code 8.3 and observe how these changes affect the results.

Answers
1. Confirm that the change in the `df` affects the width of the frequency distribution as in Fig. 8.2.
2. Combine the Codes 8.1 and 8.2 to visualize the MSD of a two-dimensional movement.
3. Change the parameter values as you want and check the consequences.

Take-Home Message
- Randomness is critical for modeling stochastic behaviors in living organisms, and we learned how to implement randomness using Python.
- Random motion was connected to diffusion, which is important to create patterns in biological systems.
- From randomness, we can obtain the Boltzmann distribution, which is useful for connecting potential energy with the probability of existence.

References

Kondo T, Kimura A. Choice between 1- and 2-furrow cytoki-nesis in Caenorhabditis elegans embryos with tripolar spindles. Mol Biol Cell. 2019;30:2065–75.

Matsumura S, Kojidani T, Kamioka Y, Uchida S, Haraguchi T, Kimura A, Toyoshima F. Interphase adhesion geometry is transmitted to an internal regulator for spindle orientation via caveolin-1. Nat Commun. 2016;7. ncomms11858

Oosawa F. [Oosawa-style, hand made Statistical Mechanics (Japanese)]. Nagoya University Press; 2011.

Théry M, Jiménez-Dalmaroni A, Racine V, Bornens M, Jülicher F. Experimental and theoretical study of mitotic spindle orientation. Nature. 2007;447:493–6.

Self-Organization of the Cell

9

Contents

> ### What You Will Learn in This Chapter
> The cell is constructed via self-organization of the internal molecular components. Various mechanisms are thought to enable this self-organization. Here, I will introduce several mechanisms known to function inside the cell, such as negative and positive feedback regulation. The established quantitative biology models include these mechanisms.

Learning Objectives

After completing this chapter, readers should be able to

1. Understand the concept of self-organization.
2. Understand the different mechanisms of self-organization working inside the cell and provide examples.

© Springer Nature Singapore Pte Ltd. 2022
A. Kimura, *Quantitative Biology*, Learning Materials in Biosciences,
https://doi.org/10.1007/978-981-16-5018-5_9

Important Concepts Discussed in This Chapter

- *Self-organization*; the appearance of a global order of a system through bottom-up, local interactions among the components.
- *Feedback regulation*; regulation in which the output of an action affects the input.
- *Symmetry breaking*; a process to create asymmetry/polarity from a symmetric state.
- *Phase separation*; spatial separation of multiple phases, while coexisting

9.1   Why Self-Organization?

As I mentioned in Chap. 2, my interest in self-organization is why I started and continue my research today. I was inspired by a book by the architect, Yoshinobu Ashihara on what he calls the hidden order in the organization of cities. The hidden order is not an order designed in a top-down manner in advance, but an order that emerges through the assembly of individual activities in a bottom-up manner without any prior design. A similar concept is generally known as self-organization. Self-organization can be observed everywhere, not only in cities or cells.

Self-organization is a central subject in biology. It has been intensively discussed in the classical book *What is life? The Physical Aspect of the Living Cell* written by Erwin Schrödinger in 1944 (Schrödinger 1944). He focused on the fact that living matter evades the decay to equilibrium (or entropy increase) and instead creates order. This book is famous for predicting a possible structure for genetic materials, which inherits order over generations, i.e., "order from order." He also emphasized the idea of order from disorder or self-organization. After the discovery of the double helix structure of DNA, the order from order aspect of life has been emphasized in molecular biology. However, many people (including me) believe both order from order and order from disorder are important. Moreover, the interplay between them is also believed to be important; however, this is a relatively uncultivated research area that requires interdisciplinary efforts between genetics and physics. I believe that the development of quantitative biology will help such interdisciplinary efforts.

9.2   Mechanisms to Create Order

In the following sections, I will introduce several mechanisms for creating order inside the cell. While I have distinguished order from disorder/self-organization from order from order, it is difficult to clearly classify whether a given mechanism is purely self-organization or not. The mechanisms introduced here enable the appearance of a global order from local interactions among the components.

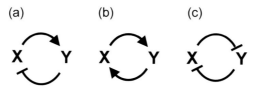

Fig. 9.1 Schemes summarizing negative and positive feedback loops. (**a**) Negative, (**b**) positive, and (**c**) double negative feedback loops

9.3   Negative Feedback Regulation

Negative feedback regulation is a type of regulation where the output of an action affects the input of the action to induce its opposite output (Fig. 9.1a). The maintenance of temperature in a room by an air conditioner is an example of negative feedback. When the air conditioner warms up the room to exceed the desired temperature, this temperature information is fed back to the air conditioner to switch off the heater or turn on the cooler. A negative feedback regulation in gene expression maintains the amount of a gene product (Ferrell et al. 2011) (see Chap. 10). If there is a time delay for the output information to be reflected in the regulation of input, negative feedback regulation can create an oscillatory pattern (Ferrell et al. 2011) (see Chap. 10). Oscillatory patterns, such as cell cycle progression, are important for establishing various rhythms inside the cell. For the modeling of negative feedback regulation, see Chap. 10 for examples.

9.4   Positive Feedback Regulation

Positive feedback regulation involves further stimulation of an action through its output (Fig. 9.1b). This is in contrast to negative feedback regulation, in which the output suppresses the action. An example of positive feedback can be seen in the growth phase of a population. An increase in the population will increase the rate of population growth as the number of parents will increase (see "logistic curve" in Chap. 6). Positive feedback amplifies small differences.

Double-negative feedback, which consists of two negative interactions, is also an example of positive feedback (Fig. 9.1c). Let us consider the following double negative feedback loop, where A represses B and B represses A. When the activity of A increases, it leads to a decrease in B, which further activates A. In contrast, if A decreases, B increases and A decreases further. This feedback regulation amplifies a small difference between A and B to make A or B dominant. We can call this switch-like behavior.

Positive feedback is used to create unidirectionality in a system. For example, during differentiation, a mechanism known as lateral inhibition is employed. Compared with its neighboring cells, an undifferentiated neuronal cell with a slightly higher amount of a particular protein (e.g., Delta) differentiates into a neuron and simultaneously inhibits the differentiation of its neighboring cells into neurons. Because these neighboring cells do not become neurons and instead differentiate into glial cells, they do not inhibit the original cell from differentiating into a neuron (Greenwald and Rubin 1992). Through this mechanism, two distinct cell types can be produced with a certain ratio and spatial pattern.

9.4.1 Positive Feedback Plus Fluctuations

Fluctuations (see Chaps. 7 and 8) play an important role in order formation involving positive feedback loops. In the lateral inhibition example introduced above, a slightly higher amount of a particular protein in a cell compared with its neighboring cells is critical for initiating specific patterning. If there was no difference in protein expression, then no pattern would appear. As discussed in previous chapters, fluctuations occur everywhere inside the cell, and thus we do not have to worry about the source of fluctuations.

The combination of positive feedback and fluctuation further creates dynamic temporal changes in patterning. As an example, I will again use our research initially introduced in Chap. 7. In cytoplasmic streaming, I mentioned that we were able to construct a model for the emergence and reversal of circulation-type streaming in *C. elegans* zygotes (Kimura et al. 2017). We proposed that the emergence of cytoplasmic streaming is due to a positive feedback loop (Fig. 9.2). Streaming is driven by motor proteins gliding along microtubules (McNally et al. 2010). The orientations of the neighboring microtubules tend to align as they are connected by the ER network. The more microtubules align, the more the flow moves in the direction of alignment, which in turn enhances microtubule alignment. This is a typical example of positive feedback. However, positive feedback alone can explain only the emergence of a collective flow but not the reversal of the flow direction. Inside the cell, microtubules appear and disappear dynamically and stochastically. This fluctuation in the lifetime of microtubules can induce a flow reversal. With the stochastic disappearance of a microtubule, alignment of the microtubules will disappear by chance every once in a while. The orientation of the next alignment that emerges is independent of the past orientation. Therefore, dynamic changes in the patterns (e.g., orientation of the cytoplasmic flow) can be induced by a combination of positive feedback regulation and fluctuation.

9.4.2 Positive Feedback Plus Negative Feedback

The combination of positive and negative feedback can generate interesting orders. An elegant example is that of the Turing patterns created by a reaction-diffusion mechanism (Kondo and Miura 2010) (Fig. 9.3). A simple version of this mechanism consists of two components, the activator (*A*) and inhibitor (*I*). *A* activates itself, leading to positive feedback. *A* also activates *I*, and *I* inhibits *A*, representing negative feedback. Another

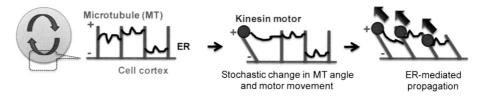

Fig. 9.2 A positive feedback model for the emergence of cytoplasmic streaming (Kimura et al. 2017)

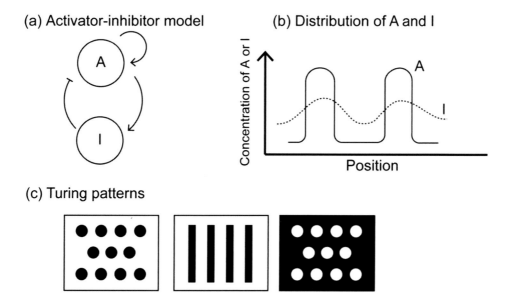

Fig. 9.3 Turing patterns made by the reaction-diffusion equation. The figure is reprinted after modification from Kimura (2019)

important component is diffusion (see Chap. 8). To create a non-uniform pattern with the reaction-diffusion mechanism, the diffusion of A should be far smaller than that of I. Once A is activated, positive feedback to increase A and negative feedback mediated by I to decrease A compete, but because the diffusion of I is large, the positive feedback will be dominant in a limited area with activated A. Meanwhile, the diffused I makes the neighboring region dominant in repressing A. Repression of A in the neighboring region reduces the production of I, and thus further regions will be A-dominant again. In this way, a striped pattern appears. We will conduct a simulation of the reaction-diffusion model in Chap. 10. This reaction diffusion mechanism was first applied to explain the striped pattern of tropical fish and was later applied to various biological patterns.

9.5   Symmetry Breaking

Enhancing small differences by positive feedback is considered an underlying mechanism of symmetry breaking in cell biology. Symmetry breaking is the process of creating asymmetry/polarity from a symmetric cell (Li and Bowerman 2010) and is considered a form of self-organization because it creates order from disorder. A mechanical process, such as cytoplasmic streaming, is known to trigger a symmetry breaking event.

In the *C. elegans* zygote, polarity of the embryo is established after fertilization but before the first cell division (Bienkowska and Cowan 2012; Kimura and Kimura 2020). Positive feedback (more precisely, double negative feedback) by mutual inhibition of the PAR-3 and PAR-2 complexes enriches PAR-3 in the cortical domain of the anterior half of the embryo and PAR-2 in the posterior half. The mechanism triggering exclusion of PAR-3 from the posterior cortex is thought to be a posterior to anterior flow at the cortical region, which is induced by the local relaxation of actomyosin (Munro et al. 2004; Mayer et al. 2010). In addition to reactions and diffusion, other mechanisms can also contribute to the creation of spatial patterns in and with the cell (Howard et al. 2011).

9.6   Phase Separation in Cell Biology

The concept of liquid-liquid phase separation (LLPS)—the coexistence of two liquid phases—is a relatively recent topic of self-organization in cell biology. The coexistence of two different phases is not difficult to imagine, as H_2O exists around us in both the liquid and vapor phase. For LLPS, one region has a higher concentration of a molecule compared with the other. As such, separation can occur spontaneously, some molecules can be concentrated in a liquid compartment without a membrane boundary in the cell. Germline P granules in the *C. elegans* zygote were the first organelles discovered to result from LLPS (Brangwynne et al. 2009). A number of cellular compartments have also been reported to be formed by LLPS; thus, LLPS is another mechanism underlying self-organization inside the cell.

Take-Home Message
- Self-organization is an important concept in biology, where order arises from a bottom-up accumulation of local interactions.
- Negative and positive feedback loops are simple but powerful mechanisms for self-organization.
- Feedback regulation alongside fluctuation and diffusion enable various types of orders to emerge or the breaking of symmetry.

References

Bienkowska D, Cowan CR. Centrosomes can initiate a polarity axis from any position within one-cell *C. elegans* embryos. Curr Biol. 2012;22:583–9.

Brangwynne CP, Eckmann CR, Courson DS, Rybarska A, Hoege C, Gharakhani J, Jülicher F, Hyman AA. Germline P granules are liquid droplets that localize by controlled dissolution/condensation. Science. 2009;324:1729–32.

Ferrell JE, Tsai TY-C, Yang Q. Modeling the cell cycle: why do certain circuits oscillate? Cell. 2011;144:874–85.

Greenwald I, Rubin GM. Making a difference: the role of cell-cell interactions in establishing separate identities for equivalent cells. Cell. 1992;68:271–81.

Howard J, Grill SW, Bois JS. Turing's next steps: the mechanochemical basis of morphogenesis. Nat Rev Mol Cell Biol. 2011;12:392–8.

Kimura A. Introduction to cell architectonics (Japanese). Kogakusha; 2019.

Kimura K, Kimura A. Cytoplasmic streaming drifts the polarity cue and enables posteriorization of the *Caenorhabditis elegans* zygote at the side opposite of sperm entry. Mol Biol Cell. 2020;31:1765–73.

Kimura K, Mamane A, Sasaki T, Sato K, Takagi J, Niwayama R, Hufnagel L, Shimamoto Y, Joanny JF, Uchida S, Kimura A. Endoplasmic-reticulum-mediated microtubule alignment governs cytoplasmic streaming. Nat Cell Biol. 2017;19:399–406.

Kondo S, Miura T. Reaction-diffusion model as a framework for understanding biological pattern formation. Science. 2010;329:1616–20.

Li R, Bowerman B. Symmetry breaking in biology. Cold Spring Harb Perspect Biol. 2010;2:a003475.

Mayer M, Depken M, Bois JS, Jülicher F, Grill SW. Anisotropies in cortical tension reveal the physical basis of polarizing cortical flows. Nature. 2010;467:617–21.

McNally KL, Martin JL, Ellefson M, McNally FJ. Kinesin-dependent transport results in polarized migration of the nucleus in oocytes and inward movement of yolk granules in meiotic embryos. Dev Biol. 2010;339:126–40.

Munro E, Nance J, Priess JR. Cortical flows powered by asymmetrical contraction transport PAR proteins to establish and maintain anterior-posterior polarity in the early *C. elegans* embryo. Dev Cell. 2004;7:413–24.

Schrödinger E. What is life? The physical aspect of the living. Cell: Cambridge University Press; 1944.

Modeling Feedback Regulations

10

Contents

What You Will Learn in This Chapter

As we learned the importance of feedback regulations for self-organization in the previous chapter, we now learn how to develop a quantitative model for the feedback regulations. We begin with modeling the feedback regulations with differential equations. Then, we develop computational codes to simulate the feedback regulations. Reaction–diffusion systems are also introduced, including both diffusion and differential equations, which we learned in the previous chapters. We also apply linear stability analyses to judge whether our models can create an interesting pattern.

Learning Objectives

After completing this chapter, readers should be able to

1. Implement feedback regulations in biological systems with python codes.

© Springer Nature Singapore Pte Ltd. 2022
A. Kimura, *Quantitative Biology*, Learning Materials in Biosciences,
https://doi.org/10.1007/978-981-16-5018-5_10

2. Apply linear stability analyses to judge whether a model described with differential equations can exhibit non-uniform patterns.

Important Concepts Discussed in This Chapter

- *Hill function*; a popular way to formulate biological reactions involving macromolecular interactions.

10.1   Basic Knowledge to Model Feedback Regulations Using Differential Equation

10.1.1 Modeling of Activation and Repression Using Hill Function

As explained in Chap. 6, differential equation is an effective means for describing the temporal changes in biological processes. A major class of such temporal changes is the change in materials, for example gene products (RNA, proteins) in a cell. The production of materials is often controlled by activators, and repressors. A popular example is transcriptional regulation. When a transcriptional activator protein binds to a regulatory DNA element, the production rate of a particular mRNA increases. In contrast, when a repressor binds to a DNA element, the production rate of mRNA is decreased. Such processes are often modeled with Hill function.

Hill function is an extended version of Michaelis-Menten equation introduced in Chap. 6 (Eq. 6.3). A.V. Hill considered the cooperativity of molecule interaction with Michaelis-Menten equation, and deduced the Hill equation (Hill 1910). In the Hill equation, the proportion of the substrate molecules bound to the enzyme is modeled as

$$nSX/X_T = S^n/(K^n + S^n) \tag{10.1}$$

Here, nSX is the concentration of X associated with n-molecules of S. n is known as the Hill coefficient. One application of the Hill equation is to model transcriptional regulations. Transcriptional regulation of gene Y by the activator or repressor X is modeled as

$$\frac{dY}{dt} = f(X) = \frac{\beta X^n}{K^n + X^n} \tag{10.2}$$

for activation, and

$$\frac{dY}{dt} = g(X) = \frac{\beta}{1 + \left(X/K\right)^n} \tag{10.3}$$

for repression reaction. Here, Y is the concentration of the gene product (mRNA), t is time, X is the concentration of activator-bound DNA element (for activation), or repressor-free element (for repression). K is a coefficient for activation or repression, β is the maximal expression level, and n is the Hill coefficient. The derivation of the Hill equation, and its application to transcriptional regulation is well explained in a textbook (Alon 2006).

10.1.2 Modeling Degradation

In the previous section, the method to model the production of a material, through activation or repression dependent on the amount of a regulator, has been explained. In the above-mentioned examples, the production rates were always positive, which implied that the quantity of the material increases. Even when the number of repressors is high, the production is low but takes a positive value. Therefore, we commonly consider the degradation of the material to model the possibility of material content decrease. A simple way to model degradation is to model the degradation rate of a material that is proportional to the amount of the product. This model assumes that each unit of the material has a constant possibility of being degraded. If the quantity of the material is large, the number of units to be degraded will be large correspondingly. This model is formalized as follows:

$$\frac{dY}{dt} = -D \times Y,\qquad(10.4)$$

where D is the degradation coefficient.

10.1.3 Negative Feedback Regulations

Let us use the model of production and degradation, to model a regulation between two components, X and Y. In this example, X activates the production rate of Y, and Y represses the production of X. This relation can be categorized into a negative feedback regulation because as X increases, Y will increase, and as a result, X will decrease. This is also true for Y because as Y increases, the production rate of Y will decrease.

Using the Hill functions and the model of degradation, the above example can be formalized as follows:

$$\frac{dY}{dt} = \frac{\beta_X X^{n_X}}{K_X^{n_X} + X^{n_X}} - D_Y Y,\qquad(10.5)$$

and

$$\frac{dX}{dt} = \frac{\beta_Y}{1 + \left(Y_{/K_Y}\right)^{n_Y}} - D_X X, \tag{10.6}$$

where K_X/K_Y, β_X/β_Y, n_X/n_Y, and D_X/D_Y are coefficients for activation/repression, the maximal expression levels, the Hill coefficients, and degradation coefficients for the regulation by X and Y, respectively.

The following Code 10.1 is a Python code to simulate the consequence of the above-mentioned model with the Euler method with time-step (dt) = 2 [unit of time].

Code 10.1 cellcycle_eular.py

```
"""
cellcycle_euler.py
Created on Wed May 12 18:05 2016
@author: akkimura (Python 3.5)
Description: cell cycle model with Euler method
"""
import numpy as np
import matplotlib.pyplot as plt

#%% Initialize parameters
beta = [1,1]
K = [0.5,0.5]
n = [2,2]
D = [0.5,0.5]
dT = 2
X0 = [1,1]
totTIME = 40
totSTEP = int(totTIME/dT)

#%% Initialize variables
deltaX = [0.0,0.0]
X = np.zeros((totSTEP+1,2))
X[0,:] = X0

#%% Calculate
for i in range(totSTEP): # i starts from 0 to totSTEP-1
    deltaX[0] = beta[1] / (1+(X[i,1]/K[1])**n[1]) - D[0]*X[i,0]
    deltaX[1] = beta[0]*(X[i,0]**n[0]) / (K[0]**n[0] + X[i,0]**n[0]) - D[1]*X
[i,1]
    X[i+1,0] = X[i,0] + deltaX[0]*dT
    X[i+1,1] = X[i,1] + deltaX[1]*dT

#%% plot results
t_values = np.linspace(0,totSTEP*dT,totSTEP+1)
```

```
for j in [0,1]:
  plt.plot(t_values, X[:,j])

# graph modifications
ax = plt.gca()
ax.set_title("Negative feedback (Euler)", size=24, weight='bold')
ax.set_xlabel("time")
ax.set_ylabel("X, Y")

#save the figure
fig = plt.gcf()
fig.canvas.get_supported_filetypes()
plt.savefig("cellcycle_euler.png")
```

The values of X and Y in the i-th step are expressed as X[i,0], and X[i,1] respectively. Similarly, K_X/K_Y, β_X/β_Y, n_X/n_Y, D_X/D_Y, and dX/dY are expressed as K[0]/K[1], beta [0]/beta[1], n[0]/n[1], D[0]/D[1], and deltaX[0]/deltaX[1], respectively. Expressing parameters for X and Y using a common name with distinct index will help in assigning the parameters and variables simultaneously. For example, K_X, and K_Y parameters are assigned in one line

```
K = [0.5,0.5]
```

, and the variables to output the results are also defined in one line.

```
X = np.zeros((totSTEP+1,2))
```

, instead of formulating two arrays (each array for X and Y, respectively).

The result of running the Code 10.1 is presented in Fig. 10.1. The values of X and Y show an oscillating behavior. Intuitively, the results appear reasonable because of the negative feedback model; when X increases, Y increases; and the increased Y represses X. As a result, the value of X decreases after X increases. Similarly, after X decreases, Y will decrease, and later, X increases. The relationship can explain an oscillatory behavior of X, and similarly for Y.

As mentioned before, it is important to estimate whether the time-step (the parameter, dt) is small enough to accurately approximate the consequence of differential equations. In fact, in the present example, if the time-step parameter is assumed to be small (e.g., dt = 1), the result will vary. Figure 10.2 is the result of the Euler method calculation when we changed dt from 2 to 1. The values of X and Y no longer show oscillatory behavior, and they plateau. As the time-step is smaller in Fig. 10.2 compared to Fig. 10.1, the former is a more accurate representation of the consequence of the differential equations (Eqs. 10.5 and 10.6). In fact, we can analytically show that Eqs. (10.5) and (10.6) do not show an oscillatory behavior.

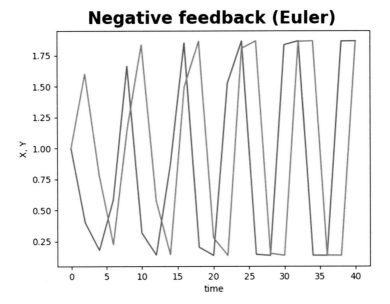

Fig. 10.1   The result of running Code 10.1 (dt = 2)

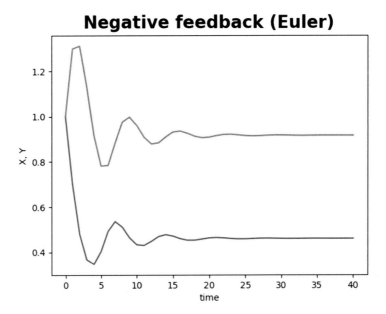

Fig. 10.2   The result of running Code 10.1 (dt = 1)

Although Code 10.1 (with dt = 2) and Fig. 10.1 were not good representations of Eqs. (10.5) and (10.6), these can be considered as appropriate models of negative feedback with a time delay. In time-delay models, the rate of production and/or degradation depends on the number of components not at that instant, but at a previous time. For example, when we consider the gene expression, it takes time from when a transcription factor binds to the promoter elements, and begins mRNA production, to the time during which the gene product (protein) is expressed, and matured to regulate a gene expression. The differential equations with time delay are expressed as follows, and Code 10.1 with large dt can be used to simulate such models.

$$\frac{dY(t)}{dt} = \frac{\beta_X X(t-\tau)^{n_X}}{K_X^{n_X} + X(t-\tau)^{n_X}} - D_Y Y(t-\tau), \tag{10.7}$$

and

$$\frac{dX(t)}{dt} = \frac{\beta_Y}{1 + \left(Y(t-\tau)/K_Y\right)^{n_Y}} - D_X X(t-\tau), \tag{10.8}$$

where τ is the time delay. (The other parameters are same as Eqs. (10.5) and (10.6)).

Questions
1. Calculate the oscillation behavior with Eqs. (10.5) and (10.6) using the Runge–Kutta method and compare the outcome with those in Figs. 10.1 and 10.2.

10.1.4 Linear Stability Analyses for Negative Feedback Models

In the above-mentioned negative feedback models, the two-component model (X and Y) without time delay converged into a set of stable values, while a similar model with time delay showed oscillation. Whether a model converges into a stable point or not can be examined with linear stabilization analyses introduced in Chap. 6.

Unlike the earlier example of logistic curve in Chap. 6, in which we considered only one component, the negative feedback model contains two components. Conducting linear stability analyses for two or more components require considerable knowledge and some techniques in linear algebra. Examples of such linear stability analyses were lucidly explained for biologists in a review article by James Ferrell Jr. and colleagues (Ferrell et al. 2011). With such analyses, we can predict how a set of differential equations behave without exploring a wide parameter space.

10.2   Reaction-Diffusion Mechanism Creating Biological Patterns

10.2.1 An Example of a Reaction-Diffusion System

A reaction-diffusion system presents an ideal example of diffusion playing a central role as introduced in Chap. 9. The mathematician Alan Turing (1912–1954) published a paper entitled "The Chemical Basis of Morphogenesis" in 1952 (Turing 1952). This paper theoretically showed that various spatial patterns, such as stripes and spots, can be created from a single mechanism, considering reaction-diffusion systems involving only two components. Patterns generated by such reaction-diffusion systems are called "Turing pattern". Later, Shigeru Kondo and his colleague demonstrated that patterns on fish skins can be explained as a reaction-diffusion system (Kondo and Asai 1995). The striking feature of this theory is that many different patterns arise without any pre-pattern.

Turing patterns can be made of two components: they are often called activators and inhibitors. The activator induces reactions to increase the amount of the activator itself and the inhibitor. The inhibitor inhibits the increase of the activator. A critical feature of the Turing model is that the inhibitor shows faster diffusion compared to the activator.

Let's see one example of a Turing pattern. Here, I introduce the Gierer-Meinhardt model (1972) (Gierer and Meinhardt 1972). The concentration of the activator (A) and inhibitor (I) will change both spatially and temporally through the following differential equations:

$$\frac{\partial A}{\partial t} = \frac{A^2}{I} - bA + D_A \nabla^2 A, \tag{10.9}$$

$$\frac{\partial I}{\partial t} = A^2 - I + D_I \nabla^2 I. \tag{10.10}$$

The first term of the right side of Eq. (10.9) means that the production rate of A is activated by A and repressed by I. We can tell that the term increases as A increases, and the term decreases as I increases. Similarly, the first term on the right side of Eq. (10.10) means that the production of I is activated by A. This model is simple compared to the Hill function to model activation and repression, introduced earlier in this chapter. Yet, it is somehow similar to the Hill function. A is squared, indicating a non-linear effect on the concentration of A to the activation of A and I. The second terms on the right sides of Eqs. (10.9) and (10.10) correspond to the degradation of A and I, as explained earlier in this chapter. The third terms correspond to the diffusion, as explained in Chap. 8.

Code 10.2 is a Python code implementing the model. The parameters for diffusion were set to 0.00001 for the activator and 0.3 for the inhibitor. The initial concentration for the activator and inhibitor were set to 2 and 4, respectively. To add noise, the concentration of the activator was set to 3 at a single point. A result of running the code is shown in Fig. 10.3. We can see 1-dimensional stripe patterns as a result.

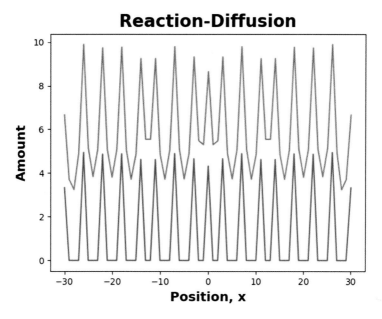

Fig. 10.3 The result of running Code 10.2

Code 10.2 ReactionDiffusion_euler.py

```
"""
cellcycle_euler.py
Created on Wed May 12 18:05 2016
@author: akkimura (Python 3.5)
Description: cell cycle model with Euler method
"""
import numpy as np
import matplotlib.pyplot as plt

#%% parameters
b = 0.5
Dif = [0.00001,0.3] # diffusion constant
dT = 0.2
x_width = 30
x_array = np.arange(-x_width, x_width+1,1)
t_step = 1000

#%% Initialize variables
deltaX = [0.0,0.0]
X = np.ones((2*x_width+1,t_step+1,2))
X[:,0,1] = 4*X[:,0,1]
X[:,0,0] = 2*X[:,0,0]
X[x_width,0,0] = 3.0 # break symmetry
```

```
#%% Calculate
for t in np.arange(t_step):
  for x in np.arange(2*x_width+1):
    # reaction
    deltaX[0] = (X[x,t,0]**2)/X[x,t,1] - b * X[x,t,0]
    deltaX[1] = (X[x,t,0]**2) - X[x,t,1]

    if x==0:
      for i in [0,1]:
      X[x,t+1,i] = X[x,t,i] + deltaX[i]*dT + Dif[i]*X[x+1,t,i] - Dif[i]*X[x,
t,i]
        if X[x,t+1,i]<0:
          X[x,t+1,i]=0
    elif x==2*x_width:
      for i in [0,1]:
      X[x,t+1,i] = X[x,t,i] + deltaX[i]*dT + Dif[i]*X[x-1,t,i] - Dif[i]*X[x,
t,i]
        if X[x,t+1,i]<0:
          X[x,t+1,i]=0
    else:
      for i in [0,1]:
        X[x,t+1,i] = X[x,t,i] + deltaX[i]*dT + Dif[i]*X[x+1,t,i] + Dif[i]*X
[x-1,t,i] - 2*Dif[i]*X[x,t,i]
        if X[x,t+1,i]<0:
          X[x,t+1,i]=0

#%% draw graph
plt.figure(1)
samplingT = t_step
plt.plot(x_array,X[:,samplingT,0])
plt.plot(x_array,X[:,samplingT,1])
ax = plt.gca()
ax.set_title("Reaction-Diffusion", size=24, weight='bold')
ax.set_xlabel("x", size=18, weight='bold')
ax.set_ylabel("amount", size=18, weight='bold')
fig = plt.gcf()
fig.canvas.get_supported_filetypes()
plt.savefig("Reaction-Diffusion.png")
```

10.2.2 Linear Stability Analysis for the Reaction-Diffusion System

Why does the reaction-diffusion system, defined by Eqs. (10.9) and (10.10), form non-uniform patterns? This can be analyzed with linear stability analysis, introduced in Chap. 6. In Chap. 6, we did not consider a case with more than one component. In this reaction-diffusion system, we have two components (A and I), providing a good example of linear stability analysis for two components.

Linear stability analysis investigates whether the equilibrium point is stable or not. Firstly, we need to obtain the equilibrium point (A^*, I^*). Eqs. (10.9) and (10.10) both consist of reaction terms ($\frac{A^2}{I} - bA$ and $A^2 - I$) and diffusion terms ($D_A \nabla^2 A$ and $D_I \nabla^2 I$). If A and I are uniform at $A = A^*$ and $I = I^*$, respectively, the concentration of A and I will not change over time, regardless of diffusion. Therefore, A^* and I^* are the solutions of the simultaneous equations:

$$\frac{(A^*)^2}{I^*} - bA^* = 0, \tag{10.11}$$

$$(A^*)^2 - I^* = 0. \tag{10.12}$$

The solution is $A^* = 1/b$, $I^* = 1/(b^2)$.

Next, let's add a small change of δA and δI to the equilibrium point of A^* and I^*, respectively. For clarity, we define $F(A,I)$ and $G(A,I)$ as

$$F(A, I) = \frac{A^2}{I} - bA + D_A \nabla^2 A, \tag{10.13}$$

and

$$G(A, I) = A^2 - I + D_I \nabla^2 I. \tag{10.14}$$

From the definition,

$$\frac{\partial (A^* + \delta A)}{\partial t} = F(A^* + \delta A, I^* + \delta I), \tag{10.15}$$

and

$$\frac{\partial (A^* + \delta A)}{\partial t} = \frac{\partial (\delta A)}{\partial t}. \tag{10.16}$$

For Eq. (10.16), note that A^* is the equilibrium point and $\frac{\partial (A^*)}{\partial t} = 0$. From the Taylor expansion of the right side of Eq. (10.15), we get

$$F(A^* + \delta A, I^* + \delta I) = F(A^*, I^*) + \frac{\partial F}{\partial A}\delta A + \frac{\partial F}{\partial I}\delta I + \dots \approx \frac{\partial F}{\partial A}\delta A + \frac{\partial F}{\partial I}\delta I. \quad (10.17)$$

Combining Eqs. (10.15)–(10.17), we can approximate

$$\frac{\partial}{\partial A}\delta A = \frac{\partial F}{\partial A}\delta A + \frac{\partial F}{\partial I}\delta I. \quad (10.18)$$

Similarly, we also get

$$\frac{\partial}{\partial I}\delta I = \frac{\partial G}{\partial A}\delta A + \frac{\partial G}{\partial I}\delta I. \quad (10.19)$$

Now, we assume

$$\delta A(x, t) = \delta A_0 \exp(\lambda t + ikx). \quad (10.20)$$

$$\delta I(x, t) = \delta I_0 \exp(\lambda t + ikx). \quad (10.21)$$

This is a mathematical technique to test whether δA and δI converge to zero after a time, determining if the equilibrium points of $A = A^*$ and $I = I^*$ are stable, or not. If $\lambda < 0$, δA and δI will converge to zero, and if $\lambda > 0$, δA and δI won't converge and the equilibrium point is unstable. From the basis of the derivative, we get

$$\frac{\partial}{\partial t}\delta A = \lambda \delta A_0 \exp(\lambda t + ikx) \quad (10.22)$$

$$\frac{\partial}{\partial t}\delta I = \lambda \delta I_0 \exp(\lambda t + ikx). \quad (10.23)$$

Putting Eqs. (10.20)–(10.23) to Eqs. (10.18)–(10.19), we obtain simultaneous equations:

$$\lambda\begin{pmatrix} \delta A_0 \\ \delta I_0 \end{pmatrix} = \begin{pmatrix} \dfrac{\partial F}{\partial A} & \dfrac{\partial F}{\partial I} \\ \dfrac{\partial G}{\partial A} & \dfrac{\partial G}{\partial I} \end{pmatrix}\begin{pmatrix} \delta A_0 \\ \delta I_0 \end{pmatrix}. \quad (10.24)$$

In linear algebra, when the relationship like Eq. (10.24) holds true, λ is called the

eigenvalue of the matrix $\begin{pmatrix} \dfrac{\partial F}{\partial A} & \dfrac{\partial F}{\partial I} \\ \dfrac{\partial G}{\partial A} & \dfrac{\partial G}{\partial I} \end{pmatrix}$, and $\begin{pmatrix} \delta A_0 \\ \delta I_0 \end{pmatrix}$ is called the eigen vector. The

condition of the equation to return a non-zero solution is

$$\det \begin{pmatrix} \dfrac{\partial F}{\partial A} - \lambda & \dfrac{\partial F}{\partial I} \\ \dfrac{\partial G}{\partial A} & \dfrac{\partial G}{\partial I} - \lambda \end{pmatrix} = \left(\dfrac{\partial F}{\partial A} - \lambda \right)\left(\dfrac{\partial G}{\partial I} - \lambda \right) - \left(\dfrac{\partial F}{\partial I} \right)\left(\dfrac{\partial G}{\partial A} \right) = 0. \quad (10.25)$$

This equation is called characteristic equation of the matrix. From Eqs. (10.13) and (10.14), we obtain the following:

$$\frac{\partial F}{\partial A} = \frac{2A}{I} - b - D_A k^2 = b - D_A k^2, \quad (10.26)$$

$$\frac{\partial F}{\partial I} = -\frac{A^2}{I^2} = -b^2, \quad (10.27)$$

$$\frac{\partial G}{\partial A} = 2A = \frac{2}{b}, \quad (10.28)$$

$$\frac{\partial G}{\partial I} = -1 - D_I k^2, \quad (10.29)$$

Putting Eqs. (10.26)–(10.29) into Eq. (10.25) yields the characteristic equation as follows:

$$\lambda^2 + \left\{ 1 - b + (D_A + D_I)k^2 \right\}\lambda + \left\{ k^4 D_A D_I + (D_A - bD_I)k^2 + b \right\} = 0. \quad (10.30)$$

To have a positive solution of λ, the following condition should be fulfilled:

$$\left\{ k^4 D_A D_I + (D_A - bD_I)k^2 + b \right\} < 0. \quad (10.31)$$

Meanwhile, to have a positive solution of k^2, the following condition should be fulfilled:

$$D_A - bD_I < 0, \text{ and } (D_A - bD_I)k^2 - 4bD_A D_I > 0. \quad (10.32)$$

The condition that fulfills in Eqs. (10.31) and (10.32) is $D_A < {\sim}0.17bD_I$. This means that the diffusion constant of the activator should be sufficiently smaller than that of the inhibitor.

In conclusion, linear stability analysis provides the condition in which the equilibrium point is unstable, and the system shows non-uniform distribution.

Questions
2. Change the parameters in Code 10.2; confirm that the parameters stabilizing the equilibrium point do not produce any patterns.

Answers
1. Please try to make a code yourself. An example of code based on the Runge-Kutta method is provided in Chap. 6 (Code 6.1).
2. For example, set the same value for D_A and D_I; e.g., Dif = [0.1,0.1].

Take-Home Message
- The Hill functions are often used to model the activation and repression of biological processes.
- As an example of creating patterns using diffusion, we learned a reaction-diffusion mechanism that creates Turing patterns.
- Linear stability analysis, which we learned in the previous chapter, was applied to a reaction-diffusion system, containing two components.

References

Alon U. An introduction to systems biology; 2006. https://doi.org/10.1201/9781420011432-9.

Ferrell JE, Tsai TY-C, Yang Q. Modeling the cell cycle: why do certain circuits oscillate? Cell. 2011;144:874–85.

Gierer A, Meinhardt H. A theory of biological pattern formation. Kybernetika. 1972;12:30–9.

Hill AV. The possible effects of the aggregation of the molecules of hemoglobin on its dissociation curves. J Physiol. 1910:4–7.

Kondo S, Asai R. A reaction–diffusion wave on the skin of the marine angelfish Pomacanthus. Nature. 1995;376:765–8.

Turing AM. The chemical basis of morphogenesis. Philos Trans R Soc B. 1952;237:37–72.

Development of the Cell over Time (Perspectives)

11

Contents

What You Will Learn in This Chapter

As discussed in the previous chapter, the self-organization of the architecture of the cell is full of mystery. Another unanswered question is how cells transition from one order to another in a reproducible manner. I call this the "development over time (problem) of the cell." A quantitative biology approach for addressing this question is to construct quantitative models for successive orders and then connect them with minimum modification between the models, or modifications supported by experimental evidence. This is a difficult challenge in modern biology, and solving this problem may pave the way to a new form of scientific research.

© Springer Nature Singapore Pte Ltd. 2022
A. Kimura, *Quantitative Biology*, Learning Materials in Biosciences,
https://doi.org/10.1007/978-981-16-5018-5_11

11.1   Development over Time: Temporal Changes from One Order to Another

Not only do cells and organisms create order in a self-organized manner, they also transition from one order to another in a dynamic and reproducible manner. Consider embryogenesis; during the early stage of embryogenesis, cell division occurs in multiple rounds. Cellular order at the interphase and mitotic phases are very different, but the cell transitions between the two phases repeatedly. Often, embryogenesis is accompanied by a reduction in cell size, and thus the transitions must be robust against changes in cell size. The later stage of embryogenesis involves complex communication between the cells and differentiation into various cell types. These are all transitions from one order to another. How this transition is achieved is a research frontier of modern biology.

11.2   An Example: Development of Cell Arrangement over Time

Early embryogenesis in animals is a good subject for elucidating time development. In many cases, gene expression from the zygotic genome is inactive, and thus changes in the gene products may not be dynamic. Changes in cell arrangement patterns accompanying cell division can be explained by relatively simple models that consider surface tension or simply the repulsion between cells (Fickentscher et al. 2013; Pierre et al. 2016).

11.3   Models for Individual but Sequential Cell Orders

In this book, I have explained how to create a quantitative model that accounts for a particular order observed in cells. One approach for understanding the development of orders over time is individually constructing models for sequential orders, and then connecting these models to explain the development over time of a cell's behavior.

I have been involved in constructing models for individual but sequential processes occurring during early embryogenesis in *C. elegans*. To date, we have constructed models for (1) cytoplasmic streaming immediately after fertilization (Kimura et al. 2017), (2) cytoplasmic streaming after symmetry breaking (Niwayama et al. 2011, 2016), (3) centration of the nucleus and the mitotic spindle (Kimura and Onami 2005, 2007; Kimura and Kimura 2011), (4) off-centration of the mitotic spindle (Kimura and Onami 2007), (5) elongation of the spindle (Hara and Kimura 2009), (6) cytokinesis (Koyama et al. 2012), and (7) cell arrangement at the 4-cell stage (Yamamoto and Kimura 2017) (Fig. 11.1). Because these processes occur sequentially, we may be able to explain the transitions if we can connect the models. However, at present, these models are based on different settings. One exception is the transition from centering to off-centering of the mitotic spindle, for which we were able to construct a unified model to explain these processes (Kimura and

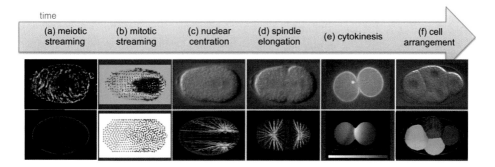

Fig. 11.1 Modeling of sequential processes during early embryogenesis of *C. elegans*. The top panels are images of real embryos, and the bottom panels show the respective quantitative models

Onami 2007). Thus, we must first unify the model settings and then determine the critical changes in the transitions from one state/process to another. The changes may reflect changes in the parameter values. Such hypotheses would be tested by manipulating the genes responsible for these parameters. Although this research is still in its infancy, we hope that by connecting each process, some general concepts regarding the transitions between cellular orders might emerge.

11.4 Transition of Different Orders: Diversity in Time Scales

The transition of orders over time is not gradual and is often drastic. How can such drastic changes be modeled with unified settings? In Chap. 7, we discuss modeling the diverse behaviors of the cell with a single model. Similarly, different orders observed during development can be explained by a unified model with different parameters. In some cases, a drastic change may be explained by a small change in the parameters of the same model. The theoretical backbone of such a transition can be provided by the physics of nonlinear dynamics, such as the idea of bifurcation (Strogatz 2001). The questions associated with cellular diversity and development over time may be connected.

11.5 Perspective

In this textbook, I have introduced some elementary concepts and programming skills for quantitative biology and focused on my own field of cell architectonics. While cell architectonics is a small field (or not even a field yet), the concepts introduced in this book, such as self-organization and diversity, are common concepts that appear in any field of biology.

Many scientists have pointed out that self-organization and related topics are new frontiers in science. Conventional science has discovered and utilized a relatively simple relationship between cause and effect. In other words, mechanisms have a "center" that provides the cause of the phenomenon. In contrast, newer scientific fields deal with phenomena that cannot be explained by a simple cause-effect relationship. In feedback regulation, an effect can also be the cause, and thus there are no distinct causes and effects. Rather than having a clear center that provides the cause, a number of components in the system can gradually (over time) provide the cause, which makes interpreting the situation difficult. Furthermore, fluctuations play important roles, and thus consequences are not deterministic.

In current scientific disciplines, these problems are addressed by nonlinear science and the science of non-equilibrium systems. Biology, as well as other scientific fields, can develop further by integrating new theories in nonlinear and non-equilibrium sciences with experimental findings. I believe that quantitative biologists will play a central role in the development of biology.

Take-Home Message
- Sequential transition from one order to another (i.e., time development) is a true frontier of modern biology.
- Quantitative modeling approaches may be an effective strategy for tackling the question of time-dependent cell development.

References

Fickentscher R, Struntz P, Weiss M. Mechanical cues in the early embryogenesis of *Caenorhabditis elegans*. Biophys J. 2013;105:1805–11.

Hara Y, Kimura A. Cell-size-dependent spindle elongation in the *Caenorhabditis elegans* early embryo. Curr Biol. 2009;19:1549–54.

Kimura K, Kimura A. Intracellular organelles mediate cytoplasmic pulling force for centrosome centration in the *Caenorhabditis elegans* early embryo. Proc Natl Acad Sci U S A. 2011;108:137–42.

Kimura A, Onami S. Computer simulations and image processing reveal length-dependent pulling force as the primary mechanism for *C. elegans* male pronuclear migration. Dev Cell. 2005;8:765–75.

Kimura A, Onami S. Local cortical pulling-force repression switches centrosomal centration and posterior displacement in *C. elegans*. J Cell Biol. 2007;179:1347–54.

Kimura K, Mamane A, Sasaki T, Sato K, Takagi J, Niwayama R, Hufnagel L, Shimamoto Y, Joanny JF, Uchida S, Kimura A. Endoplasmic-reticulum-mediated microtubule alignment governs cytoplasmic streaming. Nat Cell Biol. 2017;19:399–406.

Koyama H, Umeda T, Nakamura K, Higuchi T, Kimura A. A high-resolution shape fitting and simulation demonstrated equatorial cell surface softening during cytokinesis and its promotive role in cytokinesis. PLoS One. 2012;7:e31607.

Niwayama R, Shinohara K, Kimura A. Hydrodynamic property of the cytoplasm is sufficient to mediate cytoplasmic streaming in the *Caenorhabditis elegans* embryo. Proc Natl Acad Sci U S A. 2011;108:11900–5.

Niwayama R, Nagao H, Kitajima TS, Hufnagel L, Shinohara K, Higuchi T, Ishikawa T, Kimura A. Bayesian inference of forces causing cytoplasmic streaming in *Caenorhabditis elegans* embryos and mouse oocytes. PLoS One. 2016;11:e0159917–8.

Pierre A, Sallé J, Wühr M, Minc N. Generic theoretical models to predict division patterns of cleaving embryos. Dev Cell. 2016;39:667–82.

Strogatz SH. Nonlinear dynamics and chaos: With applications to physics, biology, chemistry, and engineering. CRC Press; 2001.

Yamamoto K, Kimura A. An asymmetric attraction model for the diversity and robustness of cell arrangement in nematodes. Development. 2017;144:4437–49.

Index

© Springer Nature Singapore Pte Ltd. 2022
A. Kimura, *Quantitative Biology*, Learning Materials in Biosciences,
https://doi.org/10.1007/978-981-16-5018-5

Printed in the United States
by Baker & Taylor Publisher Services